TABLE OF CONTENTS

Meteoric cosmogenic nuclides

"Insitu" cosmogenic nuclides

Seminars

Theses

Collaborations

Visitors

EDITORIAL

The operational agreement for the **Laboratory of Ion Beam Physics** (LIP) has been renewed by the LIP board of trustees (Schulleitung, D-PHYS, and D-ERDW of ETH Zurich, Eawag, Empa, Paul Scherrer Institut). The initial agreement signed in 2008 to establish the new organizational structure of LIP has been extended for another four years. This will enable LIP to continue its mission as a national and international center for accelerator mass spectrometry (AMS) and for the materials sciences based on ion beam technology.

It is a pleasure for us to present this summary of our year 2012 activities. It indicates the wide variety of research projects LIP members are contributing to. In addition, we thank our partners for their willingness to conduct joint research projects and to contribute so generously to this annual report.

Of particular interest are our research projects related to improvements of AMS instrumentation. Two new MICADAS systems have been completed in close collaboration with our research partners at the University of Bern and the Royal Institute of Cultural Heritage in Brussels, Belgium. Two major steps have been taken to further optimize low-energy radiocarbon detection. Firstly, the use of helium as stripper gas has been demonstrated with a significant improvement in measurements conditions. Reducing the disturbance of the ion beam phase space during the charge exchange process improves the stability of the measurement conditions which enhances the reproducibility of the isotopic ratio measurement. In addition, the overall transmission could be increased to almost 50 % due to the higher stripping yield of the 1+ charge state in helium. Secondly, we could successfully use the BerneMICADAS system to test the performance of a permanent magnet mass selective filtering stage. As energy of the order of 10-20 kW is consumed during the operation of a conventional magnetic spectrometer, substantial savings in operational costs can be realized over a time range of 15-20 years with 2000 annual operation hours. Based on these encouraging results, a new MICADAS system with only permanent magnets is in the planning stage and should be finalized in 2013. The development of the necessary permanent magnets is driven by Danysik A/S (Denmark), which started an initiative to introduce "green" magnets into accelerator and mass spectrometric systems.

It is the outstanding work of the LIP scientific, technical and administrative staff that is responsible for the success of the laboratory. In 2012, we have performed more AMS analyses than ever. We are happy with our robust and flexible financing model and are grateful for the solid base funding provided by our consortium partners.

The excellent technical infrastructure with accelerator systems ranging from 200 kV up to 6 MV terminal voltage enables us to support a vast variety of accelerator mass spectrometry techniques and ion beam application. This will continue to provide excellent service to external users and contribute significantly to the educational program of ETH.

It is our desire to continue the development of ion beam technologies and applications and to make possible a more widespread use of these powerful techniques.

Hans-Arno Synal, Peter W. Kubik

THE TANDEM AMS FACILITY

Operation of the 6 MV TANDEM accelerator

High voltage generation in a Pelletron

Application of Accelerator SIMS

OPERATION OF THE 6 MV TANDEM ACCELERATOR

Beam time statistics

Scientific and technical staff, Laboratory of ion beam physics

2012 was the first year of the 6 MV EN TANDEM accelerator being operated with the NEC Pelletron charging system that was installed in summer 2011 (Fig. 1).

Fig. 1: *The running HE chain at the terminal. Note that under running conditions the suppressor and inductor electrodes are removed when fine-tuning their position.*

Compared to the year before, the total time of operation has increased again and passed the 2000-hours mark (Fig. 2). Nevertheless, the TANDEM was not running at its full potential yet. Initial problems with the new charging system and the replacement of some terminal electronics equipment necessitated several tank openings in 2012 which required a significant part of the running time to be used for testing and conditioning ($\approx$ 20%).

The remaining active beam time was shared about equally between AMS and the materials sciences. Only three AMS nuclides are currently measured at the TANDEM, ^{10}Be, ^{26}Al and ^{36}Cl. In addition, developments at the SIMS ion source were dedicated mainly to the ions of the platinum group elements. AMS and SIMS measurements were often made at lower terminal voltages than usual because of problems with the charging system.

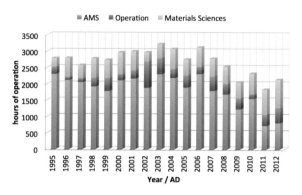

Fig. 2: *Time statistics of the TANDEM operation subdivided into AMS, materials sciences, and service and maintenance activities.*

Despite the reduced available beam time for AMS we were still able to analyze about 1000 ^{10}Be, 100 ^{26}Al and 200 ^{36}Cl samples. ^{36}Cl can only be measured in our laboratory with the higher terminal voltages of the TANDEM. Usually, ^{36}Cl^{7+} is analyzed at 6.1 MV resulting in an ion energy of about 49 MeV. Several beam times had to be performed at a lower terminal voltage (5.5 MV) or with a lower charge state (5+ instead of 7+) because foil stripping was not available at that time. However, improvements in detector design allowed sufficient ^{36}Cl-^{36}S isobar separation at the resulting lower energies.

The enhanced stability of the beam energy at low and medium terminal voltages is of benefit to various activities in the materials sciences. Besides routine ion beam analysis work and irradiations with more than 1000 samples in 2012, several new developments were realized and are documented in several contributions to this annual report.

For 2013, improvements with the Pelletron charging system are foreseen that hopefully allow operating the EN TANDEM at its full potential for routine measurements as well as for exciting new developments in AMS and the materials sciences.

HIGH VOLTAGE GENERATION IN A PELLETRON

Comparing charging efficiency of Pelletron accelerators at ETH

C. Vockenhuber, R.Gruber, S. Bühlmann

In NEC Pelletron accelerators, high voltage is generated by transporting electric charge from ground to the terminal with a rotating chain of pellets isolated from each other (Fig. 1).

The pellets are charged inductively. At ground, the pellets are in contact with the drive pulley and are charged positively using an electrode at negative voltage (called inductor). As the pellets leave the pulley (but while still inside the inductor) the charge is trapped inside the pellet and can thus be transported to the terminal. The reverse process takes place as the pellets move from the terminal to ground resulting in a doubling of the total charging current. So-called pickoff pulleys are in contact with the incoming pellets and drain some charge off the pellets providing thus high voltage on the terminal side. Suppressor electrodes suppress early discharges of the pellets as they move towards the drive and terminal pulleys. Compared to belt-operated accelerators, the Pelletron charging system has better terminal voltage stability.

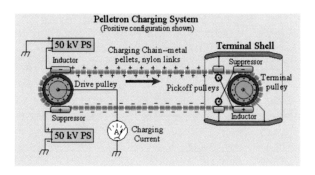

Fig. 1: *Schematics of the Pelletron charging system (www.pelletron.com/charging.htm).*

According to Q=C·V the charging current depends on the voltage and capacitance of the inductors and on the velocity of the chain nominally running at a speed of 16 m/s. Typical charging currents are 3-4 µA/kV inductor voltage. Following on an idea of the VERA AMS laboratory we implemented control of the chain

speed based on a Siemens SIMANICS G120 drive inverter. This results in improved charging stability and reduced mechanical vibrations.

At the TANDY, the rim material of the pulleys is non-conducting; lateral metal bands provide the electrical contact to the pellets. At the EN TANDEM, pulleys with conductive rim material are used eliminating the need for the lateral contact bands. This type of pulley provides a higher charging efficiency (Fig. 2) because more pellet surface is exposed to the electric field inside the inductor.

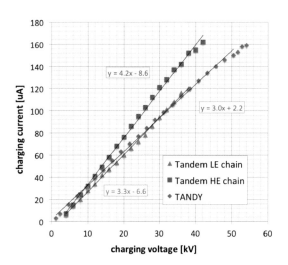

Fig. 2: *Charging efficiencies at the TANDY and EN TANDEM. Note that the TANDEM LE chain ran 7% slower than the HE chain.*

However, after about one year of operating the EN TANDEM with a Pelletron charging system it turned out that the conductive dust abraded from the chain limits the performance at terminal voltages above 5 MV. Frequent sparks occur even when the accelerator is conditioned to more than 6 MV. Similar experience at other laboratories (e.g. VERA and ANU) persuaded us to replace the TANDEM pulleys with the non-conducting version with lateral contact bands. The shutdown began in late December 2012.

APPLICATION OF ACCELERATOR SIMS

Measurement of isotopic ratios

D. Güttler, C. Vockenhuber, H.-A. Synal

At the Laboratory of Ion Beam Physics a SIMS (Secondary Ion Mass Spectrometry) ion source is coupled to the 6 MV TANDEM AMS facility. This setup combines the clean sputtering environment of the SIMS technique with the higher sensitivity of AMS which is an advantage in particular for the analysis of heavier trace elements.

Within the European "EuroGenesis" program the SIMS-AMS facility is used to analyze presolar grains extracted from meteorites. The presolar origin is revealed by isotopic ratios that differ significantly from the solar system values. Isotopic signatures of elements in the rare earth region and of platinum group elements can give hints about the processes of heavy element nucleosynthesis in the environment of massive stars and supernovae.

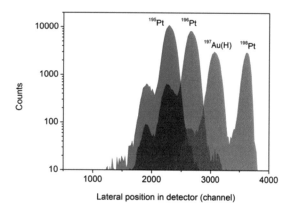

Fig. 1: *The Pt isotopes ^{195}Pt, ^{196}Pt, ^{198}Pt quasi-simultaneously injected into the accelerator are spacially separated in the position-sensitive ionization chamber.*

Reference samples with different Pt isotope ratios were used to calibrate the SIMS-AMS setup. These consisted of chemically prepared samples containing Pt isotopes in an Al_2O_3 matrix and of pure Si wafers implanted with well-defined Pt isotope doses. A background

counting rate of 1 s^{-1} was observed during the measurements corresponding to a detection limit of 2×10^{15} Pt atoms/cm^3.

For the measurement of isotopic ratios the beam bouncing system of the low-energy magnet was used to inject up to three isotopes into the TANDEM quasi-simultaneously. A wide detector entrance window allows the detection of neighboring isotopes with a relative mass difference $\Delta m/m$ of < 3 %. The position at which an isotope enters the detector can be determined with a split anode (Fig. 1). Pt- and Au-hydrides are injected together with the three atomic Pt isotopes, but the signals can be sufficiently separated from each other (Fig. 1) to accurately calculate the isotopic ratios of interest.

Fig. 2 shows measured isotopic ratios of Pt at ppm levels in a Al_2O_3 powder matrix. The measured values deviate less than 1 % from the natural ratios with a data scatter of < 4 %. The mean values are ^{196}Pt/^{198}Pt=4.71±0.02, ^{195}Pt/^{198}Pt= 3.62±0.02 and ^{195}Pt/^{196}Pt=1.30±0.01 with the error of the mean around 0.5 %.

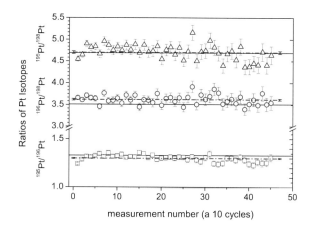

Fig. 2: *Isotopic ratios of Pt isotopes measured with Accelerator-SIMS. Black solid lines = natural ratios; dash-dotted lines = mean values.*

THE TANDY AMS FACILITY

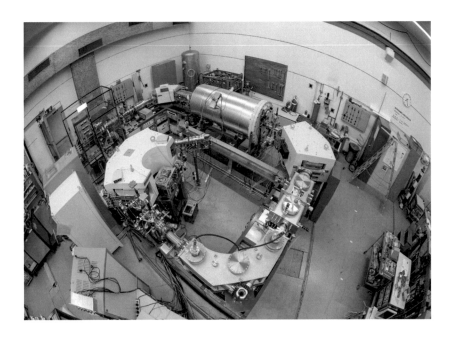

Activities on the 0.6 MV TANDY in 2012

Absorber setup for ^{10}Be detection at the TANDY

Detection of ^{26}Al^{2+}

A simplified Bragg detector

Digital pulse processing for AMS

Isotopic analysis of americium and curium

Extraction of U and Pu from seawater samples

The upgrade of the Peking University AMS

ACTIVITIES ON THE 0.6 MV TANDY IN 2012

Beam time and sample statistics

Scientific and technical staff, Laboratory of Ion Beam Physics

In 2012, the 0.6 MV multi-isotope AMS system TANDY (Fig. 1) was used for routine AMS measurements and to develop and test novel instrumentation and setups for additional AMS nuclides.

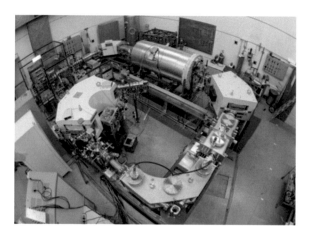

Fig. 1: *The compact 0.6 MV TANDY accelerator.*

After some major technical modifications in 2011 (e.g. the complete switchover to helium stripping), the technical focus has shifted to the development of new instrumentation. New detectors for IBA and AMS have been built and tested and new concepts for the detection of the light radionuclides ^{10}Be and ^{26}Al have been explored. Furthermore, the actinide activities of the TANDY have been expanded by developing setups for the nuclides Am and Cm.

Besides technical and instrumental develop-ments, more than 2000 hours of operation were spent on routine AMS measurements to manage the broad and diversified applications program. About half of the beam time was used for ^{10}Be and ^{129}I analyses, while 40 % was spent on both standard actinide measurements and the setup of new actinide routines (Fig. 2).

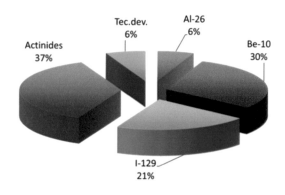

Fig. 2: *Relative distribution of the TANDY operation time for the different radionuclides and activities in 2012.*

More than 1800 routine AMS analyses of ^{10}Be, ^{129}I, and actinide samples have been performed for many different users and projects (Fig 3). ^{10}Be analyses (40% of all samples) cover ice core studies, sediment and rock samples as well as ocean water samples and manganese nodules. The remaining 60% of all samples were measured in the context of ultra-trace level studies to detect anthropogenic nuclides (^{129}I, ^{236}U, Pu, Np, Am, and Cm isotopes) in soils, in the atmosphere, in the ocean, as well as in the human body.

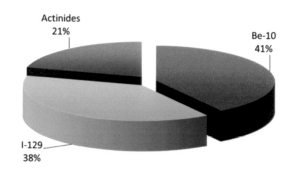

Fig.3: *Relative distribution of the number of samples measured for the various radionuclides.*

ABSORBER SETUP FOR ^{10}Be DETECTION AT THE TANDY

Setup, simulations and first measurement results

J. Lachner, M. Christl, A. Müller, M. Suter

In a modified TANDY detector setup (Fig. 1) an absorber is used to stop ^{10}B instead of using the regular degrader foil method which only reduces the ^{10}B counting rate to manageable levels for the regular detector.

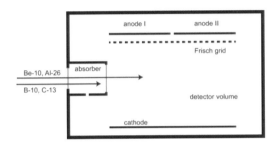

Fig. 1: *Sketch of the absorber/detector setup.*

Both volumes (absorber and detector) are filled with isobutane at the same pressure. SiN foils of different thickness separate the absorber volume from the vacuum (500 nm) and the detector (70 nm). This setup offers the potential to increase the overall efficiency of low energy ^{10}Be measurements.

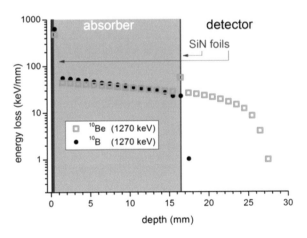

Fig. 2: *Simulations of the energy loss of ^{10}Be and ^{10}B ions with initial beam energy of 1270 keV in the absorber/detector setup.*

Helium stripping at a terminal voltage of 525 kV produces 1270 keV Be^{2+} ions with a transmission of ca. 25%. Simulations (Fig. 2) with the program SRIM-2011 [1] show that at this energy and an absorber/detector pressure of 45 mbar ^{10}B should be stopped in the absorber, while ^{10}Be should still create a signal in the detector. Measurements confirm these calculations (Fig. 3). However, ^{10}B-rich targets produce a background in the ^{10}Be gate (Fig. 3), which is not visible if the setup is run at 750 keV.

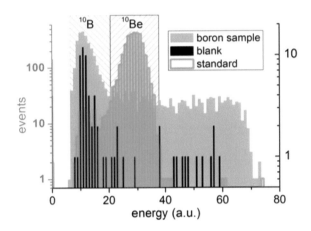

Fig. 3: *^{10}Be/^{10}B separation at 1270 keV energy.*

The high-energy background can be explained by Rutherford scattering processes of ^{10}B with H atoms in the first SiN foil. In these processes up to 420 keV are transferred to the H targets, which is sufficient for them to enter the detector volume and create a continuous background spectrum up to a fixed maximal energy. At initial beam energies of 750 keV or less, the H atoms in the initial SiN foil do not receive enough energy in the scattering process to pass through the absorber volume.

Further experiments should clarify whether a foil with lower H content or a different absorber setup can reduce the ^{10}B induced background.

[1]   www.srim.org

DETECTION OF ^{26}Al^{2+}

First results and simulations of an absorber/detector setup

J. Lachner, M. Christl, A. Müller, M. Suter

The possibility of using a modified detector setup for ^{10}Be/^{10}B separation at low ion energy via a passive absorber volume is currently being investigated [1]. In the same way, ^{26}Al measurements in charge state 2+ might become possible, as in such a detector/absorber setup ^{26}Al^{2+} could be separated from its m/q interference ^{13}C^{1+}. With He as stripper gas, Al^{2+} has the highest yield at low terminal voltages - 60 % at 0.4 MV - compared to about 15 % for the charge states 1+ and 3+.

First experiments and simulations were made to test the applicability of this detector prototype for ^{26}Al AMS at the TANDY facility (Figs. 1 and 2).

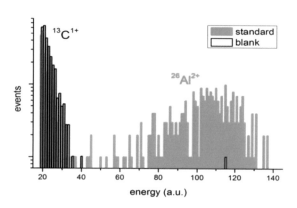

Fig. 1: *Spectra of a standard and a blank sample for an ^{26}Al beam energy of 1550 keV and an absorber/detector pressure of 22 mbar.*

The survival of molecules of mass 26 consisting of two partners of similar weight such as B, C, N, and O is not problematic - as seen for the case of ^{13}C$_2$ (Fig. 1) - because after the breakup in the entrance foil of the absorber the fragments receive only about half of the total energy and can thus not reach detector. In contrast, ^{x}Mg ions from a (^{x}MgH$_y$)$^{2+}$ breakup in the entrance foil carries most of the total energy. Due to its lower specific energy loss compared to Al it can deposit a similar amount of energy in the sensitive detector area. These troublesome

interferences can only be eliminated by increasing the stripper density to break up the molecules already in the TANDY terminal stripper. Transmission losses are the consequence of this procedure. The behavior described above can be reconstructed in detail by SRIM-2011 [2] simulations (Fig. 2).

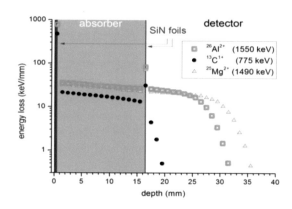

Fig. 2: *SRIM simulation for Al, C and Mg ions in the absorber/detector setup at the experimental settings applied for recording Fig. 1.*

In the tests performed, ^{26}Al standards were measured at only ca. 60 % of their nominal value. The reason for this is the non-optimal ion optics of the TANDY high-energy spectrometer for the 2+ charge state. The overall yield was thus reduced to ca. 30 % at a terminal voltage of 500 kV. Despite the losses this performance is an improvement by a factor of about two when compared to measurements in other charge states. Also, the tests were carried out with high ^{26}Al/Al ratio standards that may have caused a relatively high blank ratio on the order of 10^{-14} due to possible crosstalk in the ion source.

[1]   J. Lachner et al., Laboratory of Ion Beam Physics Annual Report (2012) 13

[2]   www.srim.org

A SIMPLIFIED BRAGG DETECTOR

Resolution measurements of a Bragg gas detector without a Frisch grid

A.M. Müller, M. Döbeli, R. Gruber

Recent experiments with simplified designs of gas ionization detectors motivated the construction of a Bragg type detector, which consists basically only of a copper plate (⌀=12 mm) mounted on a DELRIN® holder inside an aluminum tube with 40 mm outer and 25 mm inner diameter. The end flanges (one with the SiN entrance window and the other with the electrical connector) are O-ring sealed (Fig. 1).

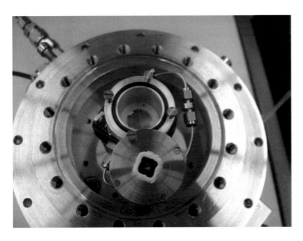

Fig. 1: *View into the open Bragg detector. The detector is mounted on a CF100 adapter flange.*

By simply shifting the anode unit along the tube the distance between anode and the detector entrance can be varied to optimize the electric field configuration. A CoolFET® preamplifier is connected to the anode.

Tests with various projectiles (He, ^{9}Be, ^{13}C, ^{27}Al and ^{35}Cl) and ion energies were performed. Bias voltage and detector pressure were first optimized with the highest measured energy for each projectile type. The ion energies were then successively reduced to measure pulse heights (Fig. 2) and resolution curves (Fig. 3).

Excellent linearity is observed between detector pulse height and energy over a large energy range even without placing a Frisch grid in front

of the anode. Compared to 'state-of-the-art' gas detectors the observed energy resolution was lower by about 10 - 30% [1].

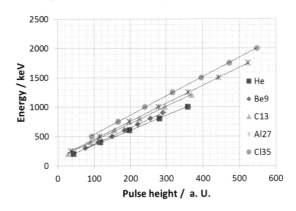

Fig. 2: *Measured detector pulse height vs. beam energy.*

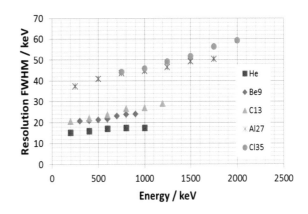

Fig. 3: *Detector energy resolution as a function of the beam energy.*

For many applications (e.g. radionuclide counting in AMS) the performance of this simple detector design should be sufficient. Future tests will show the potential for further improvements in detector resolution.

[1] A.M. Müller et al., Nucl. Instr. & Meth. B 287 (2012) 94

DIGITAL PULSE PROCESSING FOR AMS

First test of the CAEN DT5724 PHA digitizer for AMS data acquisition

A.M. Müller, N. El Hachimi[1], M. Döbeli

Various experiments [1] showed the potential of digital pulse processing as a substitute for the conventional analog data acquisition.

Recent tests were focused on the applicability and implementation of the CAEN DT5724 14 bit digitizer with 4 channels (Fig. 1) for AMS data acquisition at low energies.

Fig. 1: *CAEN DT5724 14 bit digitizer.*

The digitizer was connected to the preamplifiers of the TANDY ΔE-E_{res} gas ionization chamber (GIC) [2]. The preamplifier signals are digitized at a sampling rate of 0.1 Gs/sec and digitally filtered by converting the preamplifier pulse to a trapezoidal waveform, whose height corresponds to the number of deposited charge carriers in the detector.

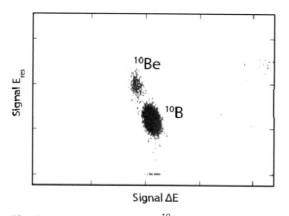

Fig. 3: *2D-Spectrum of a ^{10}Be standard sample.*

Various parameters of the digital trapezoidal filter algorithm have to be optimized to obtain the best possible performance concerning energy resolution. A program was written (in C) to control the digitizer and to access both the filter parameters and the event data (as detector pulse heights with time stamps). The program allows the identification of time correlated events in the four digitizer input channels to generate 2-dimensional ΔE-E_{res} spectra as illustrated (Fig. 2).

Ions with energies below 1 MeV produce pre-amplifier pulses with low signal to noise ratios, which seems to be a limiting factor for the actual filtering algorithm. For example, for boron at 700 keV the energy resolution was only 26 keV (Fig. 3), about 37 % higher than for analog acquisition. More work on the filtering algorithm is necessary to achieve a competitive performance in resolution for low energy ions.

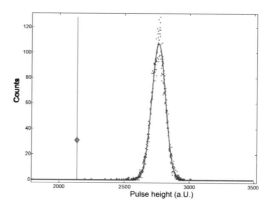

Fig. 3: *Energy spectrum of boron at 700 keV recorded with the CAEN DT5724 digitizer. The energy resolution is about 26 keV.*

[1] Laboratory of Ion Beam Physics Annual Report (2011) 34
[2] A.M. Müller et al., Nucl. Instr. & Meth. B 287 (2012) 94

[1] *École Nationale Supérieure des Ingénieurs de Caen, Caen, France*

ISOTOPIC ANALYSIS OF AMERICIUM AND CURIUM

Preliminary results of Cm/Am ratios at femtogram concentrations

X. Dai[1] , M. Christl, S. Kramer-Tremblay[1], J. Lachner, H.-A. Synal

Transplutonic elements, especially americium and curium, accumulate gradually in a reactor through continuous neutron capture reactions of uranium oxide fuel as well as tramp uranium deposits in fuel rod cladding. Due to their high radiotoxicity and intermediate half-lives, Am/Cm isotopes (e.g. ^{241}Am and $^{244/242}$Cm) are often found to be the most significant dose contributors for possible internal exposure at reactor power plants. This sometimes results in difficult issues in terms of radiological protection and monitoring, as the most commonly-used urine bioassays by alpha spectrometry do not meet the sensitivity requirements for these nuclides. On the other hand, evaluation of Cm isotopic ratios in the reactor could provide very valuable information for assessing the reactor fuel cycle characteristics and fingerprinting a particular source-term of contamination hazard as a mean of nuclear forensics [1]. All these applications would require extremely sensitive techniques (e.g. AMS) to be utilized for detection of Cm/Am isotopes.

This project aims to validate the AMS target preparation methods for Cm/Am using iron oxide or titanium oxide and to test the feasibility of the compact low energy AMS system TANDY for isotopic analysis of Cm/Am at the fg level. Good agreement between the measured ^{241}Am amounts and the expected values at the fg levels has been achieved (Fig. 1), indicating that both of the target preparation methods can be used, although the Fe oxide target is slightly favorable due to its higher signal intensity observed in the test. For both of the targets, the ionization efficiencies of Cm were found to be around three times those of Am (Fig. 2). The preliminary results demonstrated that excellent detection limits for Cm/Am isotopes (<0.1 fg) are achievable.

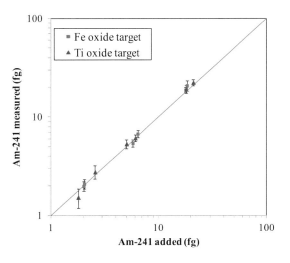

Fig. 1: *Measured vs. added ^{241}Am (fg) in the samples prepared using iron or titanium oxide targets (red line: 1:1 reference line).*

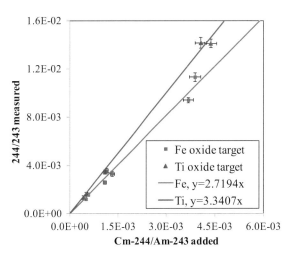

Fig. 2: *Measured vs. added atomic ratio of ^{244}Cm/^{243}Am in iron oxide and titanium oxide targets.*

[1] S. Whitney et al., Nucl. Sci. Eng. 157 (2012) 200

---

[1] *Chalk River Nuclear Laboratory, Atomic Energy of Canada Limited, Chalk River, Canada*

EXTRACTION OF U AND Pu FROM SEAWATER SAMPLES

Is there less background in charge state 3+ or 4+ at the TANDY?

A. Diebold, M. Christl, J. Lachner

Potentially, anthropogenic ^{236}U can be used as a conservative tracer to follow the spreading of water masses in the oceans [1]. The isotope system of ^{239}Pu and ^{240}Pu has already been used for this purpose [2], but since Pu is more particle reactive than U, it can also be employed to track scavenging processes.

To be able to combine both tracers, a new, efficient method for the separation of U and Pu isotopes from seawater samples was tested and then applied to eight surface samples from the North Sea and the North Atlantic Ocean (Fig. 1). After co-precipitation with Fe, the material was dissolved in nitric acid and loaded onto an UTEVA ion exchange resin. Then, Pu and U were sequentially extracted and individually prepared for AMS measurements.

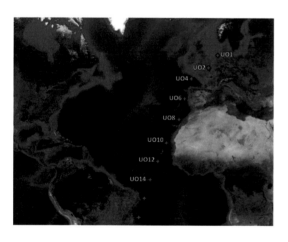

Fig. 1: *Locations and identifiers of the samples.*

Measurements of U and Pu isotopes are carried out on the TANDY AMS system in charge state 3+. As a test, U was measured in charge state 4+. The intention for this was to see whether the molecular background could be reduced compared to what has been observed in measurements with charge state 3+. Although the results agree well for both charge states (Fig. 2) strong m/q interferences from ^{118}Sn^{2+}

and ^{177}Hf^{3+} were found that make the 4+ charge state not feasible for routine measurements.

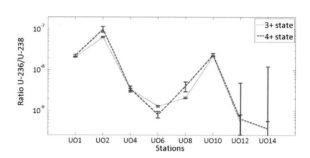

Fig. 2: $^{236}U/^{238}U$ *in dependence on the location for measurements in charge states 3+ and 4+.*

^{236}U/^{238}U ratios reached from 10^{-7} close to the nuclear reprocessing plant in the British Channel to 10^{-10} in the equatorial Atlantic (Fig. 2). In parallel ^{239}Pu (^{240}Pu) concentrations decrease from 10^7 at/kg (10^6 at/kg) in the North Sea down to 10^6 at/kg (10^5 at/kg) close to the equator (Fig. 3). The ^{240}Pu/^{239}Pu ratios are undistinguishable from those of global fallout [2].

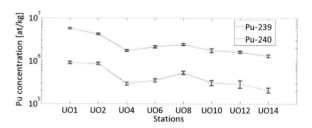

Fig. 3: ^{239}Pu *and* ^{240}Pu *concentrations in at./kg given in dependence on the location.*

In summary, the new and simple separation method works offering the integration of a Pu extraction step to prepare oceanic water samples for ^{236}U analysis in the future.

[1] M. Christl et al., Geochim. Cosmochim. Acta 77 (2012) 98

[2] P. Lindahl et al., Mar. Environ. Res. 69 (2010) 73

THE UPGRADE OF THE PEKING UNIVERSITY AMS

First ^{10}Be analyses at the 500 kV PKU AMS facility

A.M. Müller, X. Ding[1], D. Fu[1], K. Liu[1], M. Suter, H.-A. Synal, L. Zhou[2]

Recently, the Laboratory of Ion Beam Physics and Peking University (PKU) began to collaborate with the aim to modify the PKU 500 kV NEC radiocarbon facility for ^{10}Be analyses. A simple solution was found where only minor modifications of the existing instrumentation were necessary. In a first step, an ETH developed TANDY type ΔE-E$_{res}$ gas ionization chamber (GIC) was sent to Beijing to replace the existing Si-detector (Fig. 1) and a SiN degrader foil was installed in an appropriate place to reduce interfering ^{10}B currents (Fig. 2).

Fig. 1: *The ETH ΔE-E$_{res}$ GIC mounted at the beam line of the 500 kV PKU AMS facility.*

First tests with the new setup commenced in spring 2012 with very promising results. Various ^{10}Be/^{9}Be standard (Fig. 3) and blank samples were measured establishing a ^{10}Be/^{9}Be background ratio of 3.4×10^{-14} and an overall transmission of 2.2 %. This performance is already sufficient for many applications.

Future plans are to install an additional 90° magnet following the design described in [1]. This upgrade should reduce the background further and improve the overall transmission due to the energy focusing effect of the additional magnet. Our design will take up significantly less space than the present NEC

500 kV setup for ^{10}Be and might be an option for upgrading existing radiocarbon facilities.

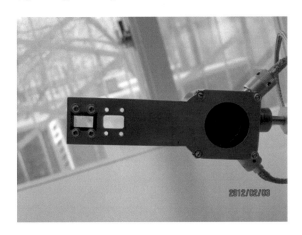

Fig. 2: *Holder for the SiN degrader foils mounted on the suppressor aperture of the ^{13}C Faraday cup.*

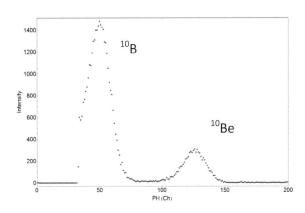

Fig. 3: *E$_{res}$ energy spectrum of a high level ^{10}Be/^{9}Be standard sample. The ^{10}Be ΔE signal was used to gate the E$_{res}$ acquisition.*

[1] A.M. Müller et al., Nucl. Instr. & Meth. B 266 (2008) 2207

^{1} *State Key Laboratory of Nuclear Physics and Technology and Institute of Heavy Ion Physics, Peking University, Beijing, China*
^{2} *Geography, Peking University, Beijing, China*

THE MICADAS AMS FACILITIES

Radiocarbon measurements on MICADAS in 2012

The myCADAS facility performance

Helium stripping of carbon ions at 45 keV

Coupling laser ablation to an AMS system

Sample preparation and measurement

CO_2 trapping for versatile AMS analysis

Straight flushing of acid released CO_2

An update on in-situ cosmogenic ^{14}C analysis

From studied object to archive

RADIOCARBON MEASUREMENTS ON MICADAS IN 2012

Performance and sample statistics

Scientific and technical staff, Laboratory of Ion Beam Physics

The MICADAS system continued routine operation last year with several capability enhancements. The software for the gas handling box was extended to permit automated measurements with both the tube cracker for glass ampoules [1] and the autosampler for direct analysis of carbonates and water samples [2] (Fig. 1).

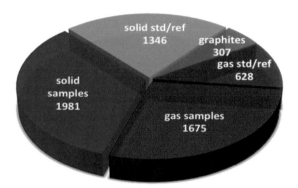

Fig. 2: *Samples and standards measured on the MICADAS prepared at ETH Zurich as solid graphite samples (blue) or received as graphite targets (gray). Samples in red were measured with the gas ion source.*

Fig. 1: *The MICADAS ion source with the gas interface and the cracker changer for the measurement of CO_2 in glass ampoules.*

The MICADAS operated on 300 days and spent 4500 hours measuring samples (Fig. 2). This was pure measurement time and more than 50% of the year! We measured more than 5900 samples and standards which was a 25 % increase over the previous year. The additional samples were produced in our sample preparation laboratory while commercial samples remained at a constant level.

The throughput of gaseous samples doubled from 742 samples (plus 212 standard and reference gases) to more than 1675 samples (plus 628 standard and reference gases). It is the first year where we have measured nearly as many gas samples as graphite.

Most of the samples were measured for our partner organizations, namely the Paul Scherrer Institute and Earth Sciences at ETH Zurich with about 900 samples each. They profited most of all from our capability to measure ultra-small samples with the gas ion source. Such small sample measurements can presently only be performed at ETH Zurich on a routine base.

For next year, some significant instrumental changes for the MICADAS system are planned again. Helium will be implemented the as stripper gas in the accelerator to gain an ~25% higher counting efficiency for ^{14}C. The new gas handling box installed at the end of the year will be in operation from the beginning of 2013 [2]. Also, a stable isotope MS (IRMS) will be added for simultaneous and precise measurements of ^{13}C and ^{14}C in gases.

[1] S. Fahrni et al., Laboratory of Ion Beam Physics Annual Report (2011) 27

[2] C. McIntyre et al., Laboratory of Ion Beam Physics Annual Report (2012) 26

THE MYCADAS FACILITY PERFORMANCE

First results from measurements at the 45 keV system

M. Seiler, S. Maxeiner, H.-A. Synal

The test setup myCADAS was built to improve the performance of radiocarbon measurements with ions of only 45 keV. The system is based on data gained with the μCADAS AMS facility [1, 2]. It is flexible enough to easily implement changes in the design (Fig. 1) such as the installation of additional apertures to remove background. Since many components are identical to those of the MICADAS system, the myCADAS setup can also be used to optimize the currently operating MICADAS radiocarbon facility.

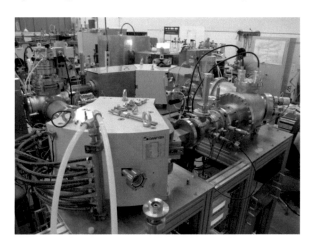

Fig. 1: *View of the myCADAS spectrometer.*

First tests showed that the beam transport through the stripper incurred large losses. The ion optical transmission through the stripper measured with negative ^{12}C ions was only slightly above 50 %, while no significant beam losses were observed through the other sections of the spectrometer. The effective beam transmission under measurement conditions (1+ charge state) was around 25 % for 15 μA of $^{12}C^-$ and even lower for higher currents. This includes the yield for the 1+ charge state and scattering losses for a gas areal density that reduces the molecular background to an insignificant level. An adjustment of the stripper position and an increase in the diameter of the

stripper tube resulted in only a slightly higher transmission but could not remove the current dependency. We found that this is caused by the ion source which produces a beam size significantly larger than simulations predicted. This problem is subject to further investigations.

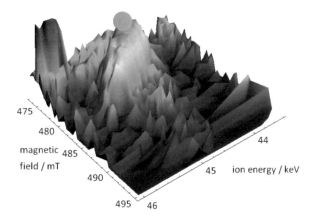

Fig. 2 *Logarithmic particle detection rate for different spectrometer settings. The green spot shows the setting for maximal ^{14}C transmission.*

The detection efficiency for radiocarbon ions could be increased to about 90 % by using an ETP AF150 electron multiplier instead of the Hamamatsu R5150-10 used with the μCADAS system.

The current background level corresponds to a radiocarbon age of about 30'000 years. An analysis of the counting rate for different spectrometer settings (Fig. 2) shows a tail of events with the same energy as the ^{14}C ions. It seems that mass separation in the magnet before the detector is not sufficient. Since the detector cannot identify the particles at these low energies, their origin is not yet known.

[1] H.-A. Synal et al., Nucl. Instr. & Meth. B 294 (2013) 349

[2] M. Seiler et al., Laboratory of Ion Beam Physics Annual Report (2011) 22

HELIUM STRIPPING OF CARBON IONS AT 45 KEV

Charge exchange and charge state yields

S. Maxeiner, M. Seiler, H.-A. Synal

The optimization of radiocarbon AMS at ion energies as low as 45 keV using helium stripping [1, 2, 3] requires knowledge of the theoretical limits of the system properties. Angular scattering, energy straggling and charge exchange in the stripper limit the maximal achievable efficiency. Selecting a high-yield charge state after the stripping process is crucial for a good overall transmission through the system. Helium has proven to be suitable concerning the above mentioned effects but its behavior at low ion energies needed to be investigated further.

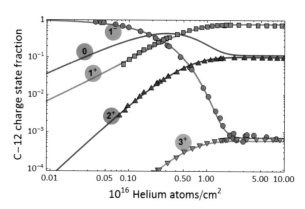

Fig. 1: *Charge state fractions of ^{12}C (corrected for losses) at different stripper densities for 50 keV ions measured at myCADAS. The curves are results of charge exchange models using fitted charge exchange cross sections.*

The evolution of different charge state fractions with gas thickness was examined with the myCADAS AMS system by varying the stripper pressure while measuring ion currents in Faraday cups before and after the stripper (Fig. 1). The measured stripper pressure was converted to stripper gas thickness by calculating the gas flow through the stripper tube. Charge state measurements were also made at the MICADAS, but at higher energies. In this case, the stripper gas density was

determined by measuring stopping power values.

The results shown in Fig. 1 can be described with simple models using single- and double-electron loss and capture cross sections. The dependency of the resulting equilibrium charge state fractions on ion energy are shown in Fig. 2 with a 1^+ charge state yield of about 75 % at 45 keV. Calculations based on fits of cross sections from other literature sources suggest even higher yields of more than 90 % at ion energies lower than 20 keV (curves in Fig. 2).

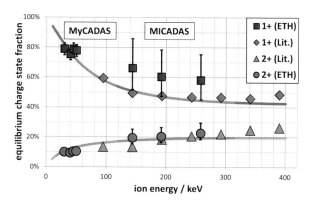

Fig. 2: *Equilibrium charge state fractions of ^{12}C in He measured at myCADAS and MICADAS combined with literature values [4, 5].*

[1]   H.-A. Synal et al., Nucl. Instr. & Meth. B 294 (2013) 349

[2]   M. Seiler et al., Laboratory of Ion Beam Physics Annual Report (2011) 22

[3]   M. Seiler et al., Laboratory of Ion Beam Physics Annual Report (2012) 23

[4]   P. Hvelplund et al., Nucl. Instr. & Meth. 101 (1972) 497

[5]   A. B. Wittkower and H.D. Betz, Atomic Data 5 (1973) 113

COUPLING LASER ABLATION TO AN AMS SYSTEM

A new cell design combines fast washout with little particle deposition

C. Münsterer[1], L. Wacker, B. Hattendorf[1], J. Koch[1], M.Christl, R. Dietiker[1], H.-A. Synal, D. Günther[1]

Laser Ablation (LA) can be used as a sampling tool to perform fast ^{14}C analysis at high spatial resolution on carbon records such as stalagmites or corals. The ablation of carbonates results in the formation of CO_2 which can be introduced directly into the gas ion source [1] of an AMS system.

For the coupling of LA with AMS a designated sample cell was constructed, which combines short washout times with the capability to host large samples, while particle deposition on cell window and walls is reduced.

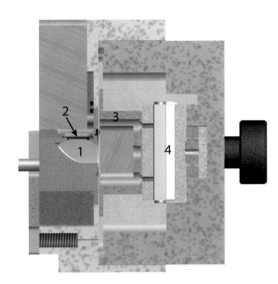

Fig. 1: *Cross-section through the cell: (1) ablation volume, (2) window, (3) sample holder, (4) positioning system.*

The cell (Fig. 1) is made of two parts to combine these features: the first part consists of the ablation region with a volume of 600 µl confining the ablation plume to a small area. This keeps dispersion small and washout quick in order to reduce cross contamination. In the second part large samples up to 150x25x15 mm^3 can be placed in a designated sample holder which keeps the sample surface at the same distance relative to the laser focus independent

of the sample thickness. A positioning system allows the precise movement of the sample underneath the laser. An ArF excimer laser is guided to and focused onto the sample through an optical setup which allows simultaneous observation.

First tests of the cell and the optical setup in combination with an ELAN 6000 Q-ICP-MS were performed on a NIST SRM 610 silicate glass. Helium with a flow rate of 1 l/min was used as a carrier gas. In a single-shot experiment the signal rise and washout of ^{27}Al were studied. The crater size was 60x400 µm for a laser energy output of 2 mJ. The results are shown in Fig. 2.

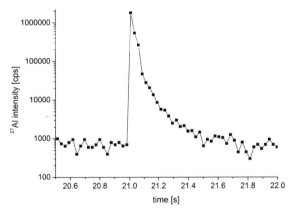

Fig. 2 *Single-shot signal of ^{27}Al from the ablation of a NIST 610 glass SRM.*

A fast signal rise is followed by a drop over 3 orders of magnitude within 0.4 seconds. This fast washout is promising for future experiments when the setup is going to be coupled to an AMS system where much lower He flow rates are used.

[1] M. Ruff et al., Radiocarbon 49 (2007) 307

---

[1] *D-CHAB, ETH Zurich*

SAMPLE PREPARATION AND MEASUREMENT

Enhancements to graphitization and gas measurement systems

C. McIntyre, L. Wacker, J. Bourquin

Apparatus used in the sample preparation laboratory and for gas measurements is continuously revised to improve efficiency and expand our capabilities. The automated graphitization equipment (AGE) [1] has been improved with new, simplified electronics and mechanical components (Fig. 1). It is routinely used for fully automated sample preparation with an elemental analyzer, carbonate system or ampoule cracker.

Fig. 1: *The next generation automated graphitization equipment (AGE3).*

The compact gas handling box and software for gas measurements has been upgraded (Fig. 2) and implemented [2, 3]. A new syringe actuator provides more robust injection of the CO_2 into the ion source. Routine measurements in an automated mode are now performed with supervision. A carbonate preparation system has been interfaced and is available for routine use.

Biogeochemical analyses require high precision ^{13}C measurements for global carbon studies. A new stable isotope MS has been selected for integration with an elemental analyzer and the gas ion source MICADAS system (Fig. 3). Automated, simultaneous ^{13}C and ^{14}C measurements of small samples such as individual compounds will be developed over next 12 months.

Fig. 2: *The latest compact gas handing box.*

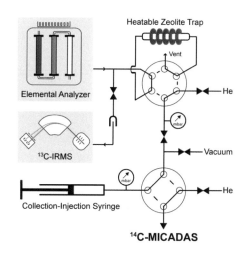

Fig. 3: *Schematic for integration of a stable isotope MS with an elemental analyzer and AMS for ^{13}C and ^{14}C measurements.*

[1] J. Rethemeyer et al., Ion Beam Physics Annual Report (2010) 36

[2] L. Wacker et al., Ion Beam Physics Annual Report (2011) 27

[3] S. Fahrni et al., Ion Beam Physics Annual Report (2011) 28

CO$_2$ TRAPPING FOR VERSATILE AMS ANALYSIS

Characterizing different CO$_2$ absorbents

G. Salazar[1], F. Rechberger, M. Seiler, C. McIntyre, L. Wacker

The gas ion source (GIS) of the MICADAS can accept CO$_2$ from a variety of analytical instruments, such as an elemental analyzer. In routine use, a gas handling box with a zeolite trap collects the CO$_2$ while the helium carrier gas is discarded. The CO$_2$ is thermally released and fed continuously into the gas ion source using a syringe [1]. Here, we compare the trap and release properties of CO$_2$ for other potential trap materials, silica gel and Carbosphere™. The materials were characterized by measuring the retention of the CO$_2$ during trap loading at various temperatures, and the temperature for complete release.

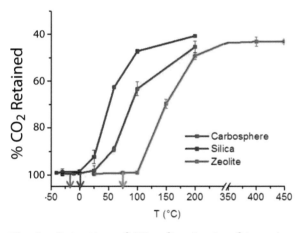

Fig. 1: *Retention of CO$_2$ after 1 min of trapping for different trap materials with increasing temperature.*

For retention, the trap was placed in a temperature controlled copper block and 2 mL of 5 % CO$_2$/He (45 μg C) mixture was loaded with 30 ml/min He. A Residual Gas Analyzer (RGA) was used to monitor the trap effluent *vs.* time. A trapping time of 1 min is normally used during measurements and Fig. 1 shows the amount of CO$_2$ retained after 1 min versus increasing temperature. The retention was practically complete (i.e. no CO$_2$ was lost) when the temperature of the zeolites was 75 °C or

lower. In comparison, the minimum usable temperature for silica gel was 0°C and -20°C for Carbosphere™.

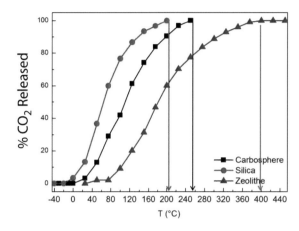

Fig. 2: *Release of CO$_2$ for different trap materials with increasing temperature.*

CO$_2$ is quantitatively desorbed from the trap at high temperatures. CO$_2$ was loaded onto the trap when cold and then heated to various temperatures. The pressure of the CO$_2$ in the syringe of the gas handling box was measured to determine the amount of CO$_2$ that was released. Fig. 2 shows how the pressure increased with temperature and reached a plateau at 100 %. Complete desorption was achieved at 400 °C for zeolites, 250 °C for Carbosphere™ and 200 °C for silica.

Silica gel and Carbosphere™ were able to effectively trap and release CO$_2$, and this occurred at lower temperatures. Zeolite remains the preferred material, however, as it allows trapping at room temperature which simplifies setup by avoiding the use of cryogenics.

[1] L. Wacker et al., Nucl. Instr. & Meth. B 294 (2013) 315

[1] *Chemistry and Biochemistry, University of Bern*

STRAIGHT FLUSHING OF ACID RELEASED CO_2

Evaluating expansive transport of sample gas to the gas ion source

M. Seiler, C. Münsterer, L. Wacker

Extracting CO_2 from carbonates by adding phosphoric acid is an efficient and fast way to prepare samples for radiocarbon dating [1]. Normally, a carrier gas like helium transports the CO_2 with a high flow rate from the reaction vessel to the gas handling system [2] where a trap separates the sample from the carrier gas again. The zeolite traps used at ETH [3] turned out to be irreproducible sources of cross contamination at the < 1 % level. To remove this problem, a setup without traps was designed to transport the CO_2 in a low helium flow by gas expansion alone (Fig. 1).

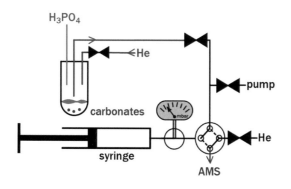

Fig. 1: *Sketch of the simplified gas handling system using expansive transport.*

The vial containing the carbonate sample is evacuated before phosphoric acid is added. The chemical reaction takes $\approx$ 5 minutes for samples containing about 50 µg C. A needle connected to the syringe feeding the gas ion source (GIS) is first sealed by inserting it into the septum of the vial so that the syringe and the connecting lines can be evacuated. Hereupon, the septum is penetrated completely to allow the gas to expand into the syringe where the pressure is measured for a rough estimate of the sample size. This result is then used to adjust the syringe volume to the optimal pressure for the GIS. Helium is then flushed through the vial into the syringe removing the remaining CO_2 in the

vial until the CO_2 is diluted to about 4 % (vol.) which leads to optimal carbon ion currents.

It is to avoid venting of the vial when penetrating the septum with one of the needles, either when adding acid or for the sampling. In the tests, the acid was added manually, while a PAL autosampler was used to evacuate and to sample. As it is crucial to reproducibly seal the needles in the septum, several needle tip styles were tested. No significant differences were observed when moving slowly into the septum.

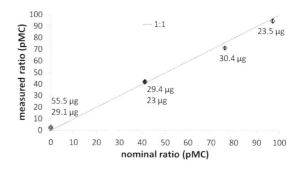

Fig. 2: *Measured sample ratios with carbon masses indicated.*

It was not possible in these tests to consistently obtain the optimal CO_2-helium mixture for the ion source because of air leaking into the vials. The carbon ion currents from the ion source were also lower than usually because the oxygen from the introduced air reduces the sputtering efficiency. However, these conditions were good enough to measure ^{14}C ratios of samples of various sizes (Fig. 2). More efforts are however needed to improve reliability and efficiency of this method.

[1] M. Schleicher et al., Radiocarbon 40 (1998) 85

[2] L. Wacker et al., Laboratory of Ion Beam Physics Annual Report (2010) 38

[3] G. Salazar et al., Laboratory of Ion Beam Physics Annual Report (2012) 27

AN UPDATE ON *IN SITU* COSMOGENIC ^{14}C ANALYSIS

Current performance of the ETH *in situ* ^{14}C extraction system

K. Hippe[1], M. Lupker[1], F. Kober[1], L. Wacker, S.M. Fahrni, S. Ivy-Ochs, R. Wieler[1]

The past few years have seen significant improvements to the analysis of *in situ* cosmogenic ^{14}C. The main difficulty is the analysis of the low *in situ* ^{14}C concentrations in terrestrial rocks without major contamination by atmospheric ^{14}C.

At the ETH extraction system (Fig. 1), *in situ* ^{14}C is routinely extracted from quartz samples and measured as CO_2 with the gas ion source of the MICADAS AMS facility. Small technical modifications of the extraction system and a shorter extraction procedure significantly improved the efficiency and reproducibility of *in situ* ^{14}C analysis [1]. Currently, the entire extraction and gas cleaning process takes about 8-10 hours and is accomplished in one day. A recent change from the previously dynamic extraction with a constant O_2 flow towards a static extraction with a fixed amount of O_2 further has further simplified the procedure.

Fig. 1: *The ETH in situ ^{14}C extraction system.*

Typical quartz samples (>10^5 ^{14}C at g^{-1}) are measured with a precision of better than 2%, while for samples with low ^{14}C concentrations (few 10^4 ^{14}C at g^{-1}) a precision of < 6% is usually achieved. Analyses of reference material

(CRONUS-A and PP-4) show a good reproducibility of 5-6% (Fig. 2) and are in reasonable agreement with published data. The blank ^{14}C contribution has been strongly reduced to currently $(3.3 \pm 1.3) \cdot 10^4$ atoms. Such low blank levels allow us to analyse quartz samples with exposure ages as low as a few hundred years [1].

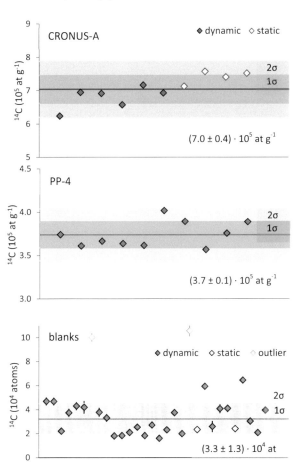

Fig. 2: *Quartz standard results (since March 2010) and blank data (since May 2011).*

[1]   K. Hippe et al., Nucl. Instr. & Meth. B 294 (2013) 81-86.

[1] *Earth Sciences, ETHZ*

FROM STUDIED OBJECT TO ARCHIVE

The path of ^{14}C samples in the preparation laboratory

I. Hajdas, C. Biechele, G. Bonani, C. McIntyre, M. Maurer, H-A. Synal, L. Wacker

In 2012, more than 2000 samples were prepared for AMS analyses in our laboratory. Each sample represents unique material, often selected from a very precious object. Preparation of material prior to ^{14}C/^{12}C measurements is a critical process that requires removal of potential contaminants and at the same time extraction of sufficient carbon from samples that are often of limited size.

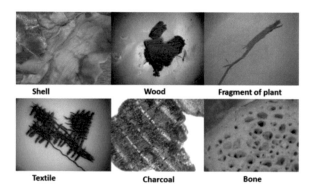

Fig. 1: *Examples of material submitted for ^{14}C dating.*

Apart from the transformation of a sample to material suitable for ^{14}C AMS analyses (graphite or CO_2) the objective of sample treatment is the separation of the original carbon from unrelated older and younger carbon that might have been incorporated into the structure of the sample. Standard treatment with physical and chemical methods removes contamination such as dirt and dust, carbonates, humic acids, and in some cases preservatives or petroleum.

Our current procedure is based on experience over the last 30 years combined with recent developments in preparation and measurement techniques. Our sample archive allows analyses of samples that were prepared and measured decades ago so that current modifications of protocols can be cross-checked [1].

Depending on sample type and preservation appropriate treatment is applied. Figure 2 shows procedures for organic matter. Bones are checked for collagen preservation and gelatine is selected and purified using ultra-filtration [2]. Carbonates are treated differently: shells, foraminifera and stalagmites with an ultrasonic H_2O bath while ostracodes, corals and some shells are leached to ca. 50 % in weak HCl [3].

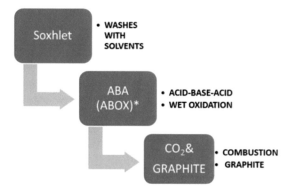

Fig. 2: *Treatment of organic matter for ^{14}C dating. Soxhlet is applied to art objects; ABA is a standard procedure; ABOX is applied to samples suspected to be older than 20 ka.*

Purified material is combusted in an elemental analyzer or dissolved in phosphoric acid and transformed to graphite using an automated graphitization system [4]. Samples that contain less than 100 μg C are only combusted and measured as CO_2 using the MICADAS gas ion source [5].

[1] I. Hajdas, Quat. Sci. Journal 57 (2010) 2
[2] I. Hajdas et al., Radiocarbon 51 (2008) 675
[3] I. Hajdas et al., Radiocarbon 46 (2004) 189
[4] L. Wacker et al., Nucl. Instr. & Meth. B 268 (2010) 931
[5] M. Ruff et al., Radiocarbon 52 (2010) 1645

RADIOCARBON APPLICATIONS

^{14}C samples in 2012

Growth rate of modern stromatolites

Eruption of the Changbaishan volcano

Discovery of a metallurgical industry in Bhutan

Pearls: Beautiful but how old?

Past hydroclimate of Inner Asia

Rock fall history of the Oeschinen area

Timeframe for paleosol architecture

Stable and labile soil organic carbon fractions

Mobilization of terrestrial carbon in the Arctic

Recent aging of Mackenzie delta lake sediments

^{14}C SAMPLES IN 2012

Overview of samples types prepared and measured at ETH

I. Hajdas, C. Biechele, G. Bonani, C. McIntyre, M. Maurer, H-A. Synal, L. Wacker

Over the last 3 years, a trend of increasing numbers of ^{14}C analyses can be observed (Tab. 1). The 2012 number of more than 5900 analyses corresponds to a 30 % increase over 2011 and includes gas target samples and samples sent to us as graphite ready for measurement. Also included are standards and blanks prepared alongside the samples.

Sample Type	2009	2010	2011	2012
Standards (OXAII)	324	290	441	615
Blanks	230	283	317	529
IAEA	85	130	130	202
Subtotal	639	703	888	1346
Archeology	518	764	669	808
Past Climate	118	191	294	275
Environment	118	145	73	30
Art	144	169	264	178
Other applications	0	26	31	246
Internal Projects	237	110	258	467
Subtotal	1135	1405	1589	2004
Gas Samples	450	993	1040	2303
External Targets		753	949	307
Grand Total	**2224**	**3854**	**4466**	**5960**

Tab. 1: *Number of samples from 2009 and 2012 prepared and measured for various applications. "Other applications" includes geochronology, carbon cycle, biogeochemistry and forensics.*

Our ^{14}C sample preparation laboratory prepared more than 2000 samples (Fig. 1) and more than 1300 standards, blanks and secondary standards (IAEA, known age material). Most of the samples submitted to us contain sufficient amounts of carbon (ca. 0.2 mg) for analysis as graphite targets. Some samples, however, contain much less and are then often prepared for measurements as gas targets. In 2012, ca. 28 samples including macrofossils, small foraminifera and fragments of charcoal or wood were dated as such.

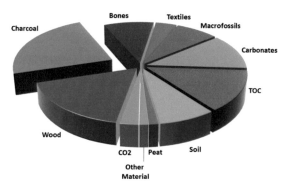

Fig. 1: *Types of sample material submitted for preparation and ^{14}C dating in 2012. "Charcoal" (23%) and "Wood" (17%) are the most common. "Carbonates" includes shells, pearls, stalagmites, foraminifera, etc. "Other Material" includes paper, antler, leather, ivory, etc.*

Similar to the years before, archeological applications and climate studies were the leading applications in 2012, although other studies saw also an increase (Tab. 1). The dating of art objects remains a small but significant application based often on longtime collaborations. Samples from Switzerland date numerous archeological excavations and archives of past environmental changes. Often, Swiss and international studies provide material from other parts of the world. Among those we studied in 2012 wood from China, shells from Iran, foraminifera from the North Atlantic, charcoal from the Chauvet cave in France and wood from a lintel of a Maya temple in Guatemala.

GROWTH RATE OF MODERN STROMATOLITES

Accretion mechanisms of stromatolites from Lagoa Salgada, Brazil

A.M. Bahniuk[1], D. Montluçon[1], T. Bontognali[1], T. Eglinton[1], C. Vasconcelos[1], I. Hajdas

Lithifying microbial mats produce laminated structures (stromatolites) that are commonly found in the geological record. Sedimentological and biogeochemical studies of modern stromatolites are of key importance for interpreting past metabolisms and paleo-environments of their ancient counterparts, which are today considered among the oldest evidence for the existence of life on Earth. This study investigates modern stromatolites from Lagoa Salgada, Brazil [1] using biomarker fingerprints to delineate paleo-environmental conditions, and radiocarbon (^{14}C) dating to determine growth rate.

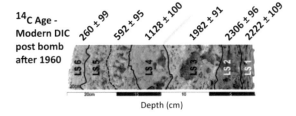

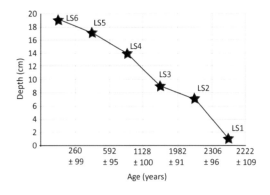

Fig. 1: *The six samples collected from different depths from the Lagoa Salgada stromatolite shows the same general age trend, with stromatolite growth initiating 2,222±109 yr. B.P. and ending with 260±99 yr. B.P.*

Combing macroscopic and microscopic sedimentological analyses it has been possible to define three major distinctive phases of stromatolitic laminae formation. The initial phase began about 2300 years before present (BP) in a water body open to the ocean with a terrestrial influence, as indicated by biomarkers trapped as intracrystalline organic matter within the stromatolite laminae.

Between 1980 and 1130 yr. BP, the stromatolites record a transitional phase wherein the water had an increased meteoric signal but $\delta^{13}C$ values of inorganic carbon became strongly enriched. The final phase,

dated between 592 and 200 yr. BP, represents a fully microbial influenced system with Mg-calcite, Ca-dolomite and authigenic clays precipitating to form microbialite laminae.

Samples	Fm	Error (%)	Age (yr B.P.)	Error (yr)	$\delta^{13}C$ (‰)
LS1	0.7583	1.36	2.222	109	-2.1
LS2	0.7505	1.19	2.306	96	9.4
LS3	0.7813	1.14	1.982	91	15.1
LS4	0.8690	1.24	1.128	100	20.5
LS5	0.9290	1.18	592	95	14.8
LS6	0.9681	1.23	260	99	8.2

Tab. 1: *Radiocarbon data for the 6 samples.*

The overall growth rate occurring at Earth's surface condition is 1 cm/100 a. This finding provides constraints on the formation rate of stromatolitic successions present in the geologic record.

Fig. 2: *Heterogeneous growth rate for the Lagoa Salgada stromatolite.*

[1] B. Turcq et al., in: Environm. Geochem. of Coastal Lagoon Syst. 6, Série Geoquímica Ambiental, eds. B. Knoppers, E.D. Bidone, J.J. Abrão, Rio de Janeiro, Brazil (1999) 25

[1] Geology, ETH Zurich

ERUPTION OF THE CHANGBAISHAN VOLCANO

^{14}C wiggle match dating of a 264 year old tree trunk (with bark).

J. Xu[1], I. Hajdas, T. Liu[2], B. Pan[1]

The 10th century A.D. eruption (or so-called Millennium eruption) of the Changbaishan volcano at the border between China and North Korea was one of the biggest in recorded history (volcanic explosivity index: VEI 7). The exact timing of the event, however, has been under intense debate for nearly three decades and no unambiguous and consensus radiometric chronology exists at present.

We collected a 264 year old larch tree (*Larix*) log which still had its bark. It was about 3 m long and 0.73 m in diameter, and only partially charred. It was buried in ≈ 14 m thick pyroclastic flow deposits at Xiaoshahe on the west slope of the Changbaishan volcano (Fig. 1). The tree was buried during the Millennium eruption and thus be used to construct a chronology.

Fig. 1: *A 264 year old larch tree log buried within pyroclastic flow deposits at site Xiaoshahe.*

The tree rings were counted inwards from the outermost ring (R1) to the pith (R264) using high-resolution images of a sawed and polished disk of the tree log. No pseudo or missing rings were found in the sequence. Annual ring samples (R1, R10, R20, … R260) were cut from the polished tree slab at 9- or 10-ring intervals (Fig. 2).

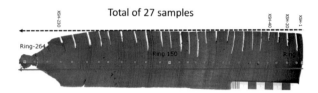

Fig. 2: *Polished tree slab for sample collection.*

In order to build a high-precision ^{14}C chronology for the Millennium eruption, we adopted a Bayesian modeling approach using the program OxCal v.4.1.7 (https://c14.arch.ox.ac.uk).

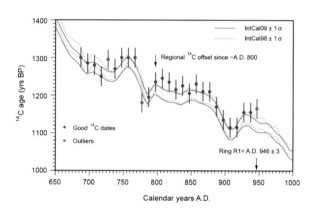

Fig. 3: *^{14}C wiggle matching plot showing the best-fit date of A.D. 946 ± 3 for the Millennium eruption.*

As a result (Fig. 3), we obtained a well-constrained calendar age of A.D. 946 ± 3 [1] that perfectly matches the inferred historical date of A.D. 946 for the Millennium eruption of the Changbaishan volcano.

[1]   J. Xu et al., Geophys. Res. Lett. 40 (2013) 1

[1] *Geology, China Earthquake Administration, Beijing, China*
[2] *LDEO, Columbia University, Palisades, USA*

DISCOVERY OF A METALLURGICAL INDUSTRY IN BHUTAN

Chemico-mineralogical analysis and ^{14}C-dating of slags

D. Oppler[1], I. Hajdas, R. Soulignac[2], V. Serneels[2]

The archaeological excavations of the University of Basel in 2009 on a 15th to 16th century monastery-castle in Bhutan, so-called Drapham-Dzong, brought some iron slags to light. Moreover a smelting place and a huge slag heap were found in the bamboo jungle close to the castle. This raised the issue, whether the iron smelting stands in context with the wealth of the monastery-castle.

Fig. 1: *The monastery-castle Drapham-Dzong.*

Slag samples were analyzed at the Laboratory of Geoscience, University of Fribourg, with X-Ray Fluorescence (XRF) and XRay Diffraction (XRD). Well preserved charcoal found while cutting some slag samples enabled ^{14}C dating.

The chemical and mineralogical analyses revealed that the slags were smelted at temperatures of a bloomery furnace (approx. 1'250 °C) and that wood was used as fuel. Only iron was smelted, but the associated mineral used could not be elicited. The chemico-mineralogical composition provides only that a granitic, regional hematite-containing mineral could have been used.

A sample, which originates from the castle, could be radiocarbon dated to the 15th to 16th century. This dating is consistent with the ^{14}C analyses of excavated ceramic samples.

However, the charcoals of the smelting place date back to the 12th to 13th century AD. It was not possible so far to prove whether this time difference is in accordance with the archeological facts or whether this is an example of a prominent 'Old wood effect', i.e. approximately 300 years old wood was used to heat the furnace, which would certainly be possible in an area of dense and uncultivated jungle.

H107a

Fig. 2: *Cut bloomery furnace slag with well-preserved charcoal.*

It has to be kept in mind, that only stray finds of a brief survey were analyzed, without vertical or horizontal stratigraphical context. For this reason, the sampling cannot be considered representative but indicates somewhat the first discovery of a metallurgical industry at the foot of the Himalayas. We hope to go back in the near future and to obtain the authorization for an archaeological excavation of this place.

[1] *Archaeogenetics, University of Basel*
[2] *Geosciences, University of Fribourg*

PEARLS: BEAUTIFUL BUT HOW OLD?

^{14}C dating as a new approach for pearl testing and identification

M.S. Krzemnicki [1], I. Hajdas

Due to their beauty, pearls have been used for adornment since pre-historic times and are among the most priced jewels, as they are connoted not only with beauty and rarity, but also with status. They have been used as means of representation in many cultures since ancient times. Famous historic jewels and ornaments with pearls are known from the treasures of the royal courts in Europe, Russia, the Middle East, India, and China. With the development of pearl cultivation in the beginning of the 20th century, pearls became more common than in previous ages.

Fig. 1: *La Peregrina pearl, a historic natural pearl found in the 16th century (inset: painting by Hans Eworth (ca. 1520-1574 AD) of Queen Mary I of England and Ireland, wearing a pendant with the La Peregrina pearl.*

Nowadays the pearl trade is a multi-billion share of the worldwide jewellery market. Its products range from low-quality and inexpensive freshwater cultured pearls to rare and highly sought-after natural pearls of historic provenance, such as the *La Peregrina* pearl (Fig. 1).

Pearls are calcium carbonate ($CaCO_3$) concretions, formed through biomineralization by both freshwater and saltwater molluscs. Such sources of carbon imply the need of reservoir correction for radiocarbon ages of pearls (Fig. 2).

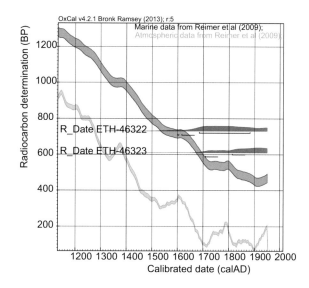

Fig. 2: *Calibrated ages of two historic pearls that were formed before the bomb peak (1950 AD). Both pearls originate from the Arabian Gulf. The ^{14}C ages were calibrated using the marine calibration curve (blue) INTCAL09 [2].*

Nevertheless, ^{14}C analysis can be a useful tool in pearl identification together with commonly used physical tools such as: radiography, X-ray luminescence, X-ray computed muon tomography and others as well as meticulous microscopic examinations.

[1]   M. Krzemnicki and I. Hajdas, Radiocarbon submitted
[2]   P. Reimer et al., Radiocarbon 51 (2009) 1111

[1] *Swiss Gemmological Institute SSEF, Basel*

PAST HYDROCLIMATE OF INNER ASIA

Reconstructing past moisture fluctuations in the Tarim Basin, China

A.E. Putnam et al.[1], I. Hajdas

The response of Central Asian water resources to climate change is uncertain, posing a major challenge to 21st Century policy and planning. Here, we approach this problem from a palaeoclimatic perspective. We combine field observations from the Tarim Basin with dendrochronology and radiocarbon dating to reconstruct how Asian water availability has fluctuated with past climate change.

During the summers of 2010 and 2011, A.E. Putnam led field teams to the remote Tarim Basin of western China to investigate the glacial and hydrological history of Central Asia during and since the last ice age. These expeditions led to the discovery of widespread stands of ancient poplar and tamarisk tree stumps that occur rooted in water-lain silt beds deep within the Taklamakan Desert, a hyper-arid region now dominated by migrating sand dunes (Fig. 1).

Fig. 1: *Sub-fossil* Populus euphratica *stumps associated with waterlain silts in the Taklamakan Desert of the Tarim Basin.*

We also found fresh-water gastropod shells within silt beds associated with now-dry distributary channels of the Tarim River, as well as mollusc shells associated with high shorelines of the now-dry lake Lop Nor (Fig. 2).

Fig. 2: *Lop Nor playa near an elevated shoreline. Mollusc shells were collected for* [14]*C dating.*

Dendrochronogical work took place at the Lamont-Doherty Earth Observatory Tree Ring Laboratory. Radiocarbon dating of wood fragments, tree rings, and shells was conducted at the AMS facilities at ETH and at the University of California, Irvine.

Ages assayed from samples collected at these distant sites, spaced over ~800 km in the Tarim Basin, afford coherent ages. Our data point to the ability of the hydroclimate of the great Inner Asian deserts to undergo rapid changes in response to distant climate anomalies.

[1]*A.E. Putnam, W.S. Broecker, E.R. Cook, J. Martin and L. Andreu Hayles, LDEO, Columbia University, Palisades, USA; D.E. Putnam and C. Wang, Environmental Studies and Sustainability, University of Maine, Presque Isle, USA; G.H. Denton, Geology, University of Maine, Orono, USA; J.D. Palmer, Gondwana Tree-Ring Laboratory, New Zealand; J. Southon, KCCAMS Facility, University of California, Irvine, USA; P. Quesada, Fore River Company, Portland, USA; F. Quesada, COL, Harvard University, Cambridge, USA*

ROCK FALL HISTORY OF THE OESCHINEN AREA

Radiocarbon-dated lake sediments as archive of past rock fall activities

S. Knapp[1], A. Gilli[1], I. Hajdas, P. Köpfli[1], J.R. Moore[1], S. Ivy-Ochs

Understanding the temporal and spatial occurrence of mass movements is of crucial interest for implementing measures to prevent and mitigate this natural hazard in the Alpine realm. An area largely impacted by the past occurrence of rock falls is the circular deep valley around Lake Oeschinen near the village of Kandersteg in the central Swiss Alps.

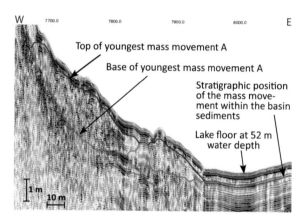

Fig. 1: *Seismic profile through the youngest observed mass movement identified by its chaotic-to-transparent seismic facies.*

In order to reconstruct past rock fall activities, we followed an approach proposed by Deplazes et al. [1] and thoroughly investigated Lake Oeschinen's subsurface using a high-resolution 3.5-kHz seismic pinger device (Fig. 1). The detailed analysis of the seismic profiles revealed the presence of eight sedimentary bodies of chaotic-to-transparent seismic facies, which are interpreted as mass movement deposits. Based on the spatial distribution of these mass movement deposits within the lake basin, the deposits can be tentatively related to observed head scarps and sliding planes around Lake Oeschinen.

A subsequent coring campaign was successful in recovering the entire, over 14.6 m-thick sedimentary succession down to the onset of

the lacustrine phase (Fig. 2), as the lake is dammed by a large rock avalanche that occurred in the mid/late Holocene [2]. Several small wood pieces found in the sediment core were radiocarbon-dated and allow establishing a chronology of the core. By linking the stratigraphic position of different mass movement deposits on the seismic profile to the coring location an age for each mass movement can be assigned. This results in a rock fall history of the area and provides the base for further hazard assessment.

Fig. 2: *Coring platform with head scarps and sliding surfaces of several mass movement events in the background.*

[1]   G. Deplazes et al. Terra Nova 19 (2007) 252
[2]   P. Köpfli et al., Laboratory of Ion Beam Physics Annual Report (2012) 57

___

[1] *Geology, ETH Zurich*

TIMEFRAME FOR PALEOSOL ARCHITECTURE

A late Quaternary basin sequence from Bologna, Northern Italy

A. Amorosi[1], L. Bruno[1], V. Rossi[1], P. Severi[2], I. Hajdas

Paleosol stratigraphy is a technique commonly applied in basin-margin position to depict cyclic architecture of late Quaternary alluvial deposits. The town of Bologna, located at the southern margin of the Po Basin, in a slightly elevated, triangle-shaped zone comprised between two adjacent fluvial systems (Reno River to the west and Savena River to the east) is an excellent site to perform stratigraphic investigation with high spatial and temporal resolution. This area, for which conspicuous background stratigraphic information is available [1], hosts an abundance of stratigraphic data stored in the Emilia-Romagna Geological Survey database (about 3,000 stratigraphic logs).

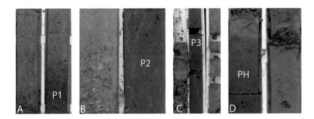

Fig. 1: *Representative core photographs of Late Pleistocene paleosols. Core bottom is lower left corner. Core width in the photographs is 10 cm.*

The recent drilling of 12 continuous cores as part of the project for the Line 1 of the Bologna Underground yielded high-quality material for high-resolution stratigraphic correlations across the Bologna interfluve (Fig. 1). Core examination revealed the presence of thin, repeated paleosol sequences as the only stratigraphic marker within apparently homogeneous, mud-prone deposits. The chronological framework was ensured by radiocarbon dating of 29 samples (20 with AMS). Based on these dates, three weakly developed paleosols (P1-P3), are of late Pleistocene age (between 40 and 25 ka). This prolonged phase of soil development is inferred

to correlate with the final stage of the stepwise late Pleistocene sea-level fall, culminating in the marine isotope stage 3/2 transition. A fourth laterally extensive Inceptisol (Fig. 2), encompassing the Pleistocene-Holocene boundary (PH), represents the major phase of soil development since the Last Glacial Maximum and appears to be related to fluvial terrace formation during the Younger Dryas.

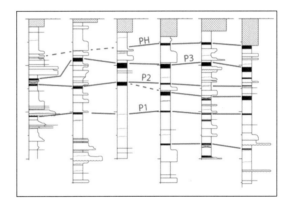

Fig. 2: *Lateral tracking of paleosols P1, P2, P3 and PH along a 3 km-long transect. The thick 'A' horizon of paleosol P3 (in black) is the most prominent stratigraphic marker across the study area.*

By contrast Holocene paleosols (with the exception of the Iron Age-Roman paleosol, which reflects a predominantly anthropogenic control) appear laterally discontinuous and invariably more immature (Entisols) than Pleistocene paleosols.

[1]   A. Amorosi et al., Sedim. Geol. 102 (1996) 275

[1] *Biological, Geological and Environmental Sciences, University of Bologna, Italy*
[2] *Geological, Seismic and Soil Survey of Regione Emilia-Romagna, Bologna, Italy*

STABLE AND LABILE SOIL ORGANIC CARBON FRACTIONS

Do permafrost conditions affect the age of C fractions in Alpine soils?

B. Pichler[1], D. Brandová[1], S. Ivy-Ochs, C. Kneisel[2], M. Egli[1]

Permafrost is a widespread phenomenon in cold alpine and subarctic environments. It is generally assumed that permafrost soils store a considerable amount of organic carbon, even though possible mechanisms of carbon stabilization by processes other than freezing are still unknown [1, 2]. Only little data is available about the storage of organic carbon, its chemical characteristics and turnover times in permafrost soils of the Alps. Consequently, the present work focused on this issue by comparing permafrost soils with nearby non-permafrost soils in alpine and subalpine climate ranges in the Swiss Alps (Upper Engadine). We hypothesize that permafrost soils differ distinctly both in their soil organic carbon (SOC) characteristics and mean SOC residence time.

One sub-Alpine and two Alpine sites were investigated. All of them had sub-sites (3 soil profiles each) that were either or not influenced by permafrost. The soil organic matter (SOM) fraction (bulk, labile and stable C) was analyzed in more detail. Our conceptual approach was based on the finding that partial oxidative degradation (using H_2O_2) of organic matter (OM) leaves behind intrinsically resistant as well as mineral-protected organic materials.

Surprisingly, the Alpine grassland and sub-Alpine forest sites showed similar organic carbon stocks. A higher primary productivity and consequently higher litter production and C inputs in the root zone were expected for the forested sites. The lack of difference is most likely due to a greater carbon input at high altitudes during a longer period of soil evolution due to a warmer climate (Atlantikum, some 5000 to 8000 years ago) and a different vegetation type.

Above the timberline, the bulk SOC showed a distinctly lower pMC at permafrost sites which was even more evident for the stable carbon

fraction (resistant to H_2O_2 treatment), where ages of up to 11 ka were recorded. Consequently, climatic conditions and the occurrence of discontinuous permafrost resulted in a very low turnover rate of SOM. At the sub-Alpine site, the differences between permafrost and non-permafrost sites were smaller or did not exist.

Alpine soils had a complex soil genesis that affects their present-day and future behavior (in light of climate change). Most likely, carbon accumulation and carbon fractionation occurred over a rather longer period (c. 5 – 8 ka BP) in a warmer climate that still imprints soil characteristics to a certain degree.

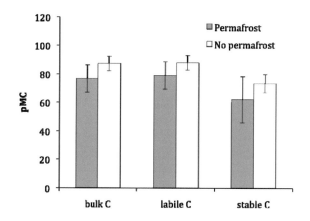

Fig. 1: *pMC values of the bulk, labile and stable SOC in high Alpine soils with and without permafrost.*

[1]   G. Hugelius and P. Kuhry, Global Biogeochem. Cycles 23 (2009) GB3006
[2]   M.W.I. Schmidt et al., Nature 478 (2011) 49

[1] *Geography, University of Zurich*
[2] *Geography and Geology, University of Würzburg, Germany*

MOBILIZATION OF TERRESTRIAL CARBON IN THE ARCTIC

Perspectives from lignin phenol ^{14}C ages

X. Feng[1], J.E. Vonk, B.E. van Dongen[2], Ö. Gustafsson[3], I.P. Semiletov[4,5], L. Wacker, T.I. Eglinton[1]

Mobilization of the large reservoirs of Arctic permafrost organic carbon (OC) constitutes a pressing concern in climate change and remains a challenge to study due to the highly diverse carbon pools associated with various physio-geographic regimes [1-2]. By radiocarbon dating specific groups of terrestrial molecular markers in river-integrated estuarine sediments across the climosequence of the Eurasian Arctic (Fig. 1), this study reveals contrasting ^{14}C character-istics and mobilization mechanisms for different carbon pools associated with Arctic fluvial transfer.

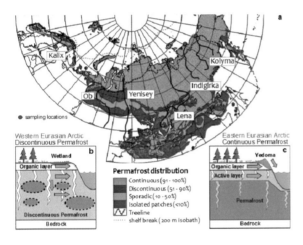

Fig. 1: *The Eurasian Arctic transect and cartoon of hydrological mobilization of terrestrial carbon into rivers. Blue arrows indicate hydrological transport of carbon.*

Briefly, while vascular-plant-derived lignin phenols incorporate significant inputs of young carbon from surface sources, plant wax lipids predominantly trace highly aged OC mobilized from deeper permafrost soil (Fig. 2). Using an isotope mixing model for plant-derived OC (represented by hydroxy phenols from both vascular plants and mosses), we estimate about equal contributions from surface organic layers and deep permafrost to sedimentary OC supplied to the East Siberian estuaries.

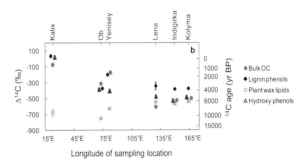

Fig. 2: *Contrasting radiocarbon contents of terrestrial markers as compared with bulk organic carbon (OC) in the estuarine surface sediments. Plant wax lipids constitute n-alkanes (C$_{27,29,31}$) and n-alkanoic acids (C$_{24,26,28}$).*

Moreover, while deep carbon mobilization is mainly controlled by permafrost distribution, young surface OC is more efficiently delivered to estuaries in high-runoff systems. As runoff is projected to increase in the future Arctic [3], anticipated signs of old permafrost OC exhumation and export may be masked by increased transport of surface detrital carbon.

[1] C. Tarnocai et al., Global Biogeochem. Cy. 23, (2009) GB2023

[2] L. Schirrmeister et al., J. Geophys. Res. 116, (2011) G00M02

[3] S. Pohl et al., Arctic 60 (2007) 173

[1] *Also, Marine Chemistry & Geochemistry, Wood Hole Oceanographic Institution, USA*
[2] *Atmospheric and Environmental Sciences (SEAES), University of Manchester, UK*
[3] *Applied Environmental Science (ITM), Stockholm University, Sweden*
[4] *International Arctic Research Center (IARC), University of Alaska, Fairbanks, USA*
[5] *Pacific Oceanological Inst., Russian Academy of Sciences, Vladivostok, Russia*

RECENT AGING OF MACKENZIE DELTA LAKE SEDIMENTS

[14]C analyses on leaf waxes and bulk OC in a 120-yr sediment record

J.E. Vonk[1], L. Giosan[2], A. Dickens[3], C. McIntyre, L. Wacker, I. Hajdas, D. Montlucon[1], T.I. Eglinton[1]

Arctic permafrost holds about half of all belowground soil organic carbon (OC) on our planet [1]. Under intensifying climate warming, these frozen OC pools may thaw and remobilize their OC, for example through fluvial transport. The Mackenzie River in Arctic Canada is the largest source of particulate OC to the Arctic Ocean [2]. Half of the Canadian permafrost is within 2 degrees of thawing, and the active layer deepening in the Mackenzie basin shows a significant trend in the last 25 years [3].

The Mackenzie River delta receives a large amount of fluvial sediments, which are stored in (annual) layers in the thousands of small and shallow delta lake. These lake sediments are valuable archives to provide context for ongoing climate-induced changes in the drainage basin.

Lake 7, a low-closure lake in the middle-delta, shows continuous, ca. 1 cm thick yearly laminations spanning the last ca. 90 yrs. Down-core bulk OC Δ^{14}C signals are between -800 and -900 ‰ (Fig. 1), suggesting a significant contribution of ancient OC from sedimentary rocks. Compound-specific ^{14}C measurements on methylated long-chain *n*-alkanoic acids from leaf waxes showed more enriched Δ^{14}C signals between -700 and -450‰ (Fig. 1). This suggests an increasing flux of aged terrestrial OC during the last ~15 years, potentially originating from remobilized permafrost.

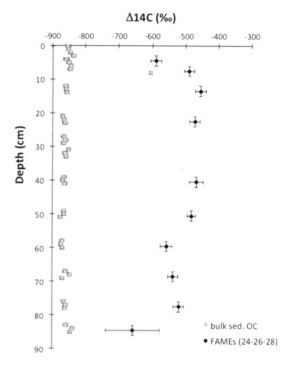

Fig. 1: *Δ^{14}C on bulk sediment OC (grey) and fatty acid methyl-esters ($\sum C_{24}-C_{26}-C_{28}$; black) in a sediment core from Lake 7, Mackenzie Delta.*

[1] C. Tarnocai et al., Global Biogeochem. Cycles 23 (2009) GB2023
[2] R.M. Holmes et al., Global Biogeochem. Cycles 16 (2002) 1098
[3] C. Oelke et al., Geophys. Res. Lett. 31 (2004) L07298

[1] *Geology, ETH Zurich*
[2] *Geology & Geophysics, Woods Hole Oceanographic Institution, USA*
[3] *Mt. Holyoke College, South Hadley, USA*

METEORIC COSMOGENIC NUCLIDES

http://www.geology-israel.co.il/WEB%20PAGE/GEO1-1-25=1.HTML

Estimation of erosion rates using meteoric ^{10}Be

^{10}Be in Swiss lacustrine sediments

^{10}Be and the Laschamp event in sediments

ESTIMATION OF EROSION RATES USING METEORIC ^{10}Be

Alpine soils under permafrost and non-permafrost conditions

B. Pichler[1], D. Brandová[1], P.W. Kubik, S. Ivy-Ochs, C. Alewell[2], C. Kneisel[3], M. Egli[1]

Permafrost ecosystems are highly sensitive to climate warming. The expected changes in the thermal and hydrological soil regime might have crucial consequences on soil erosion processes. The analysis of geomorphologic processes since the beginning of soil formation can provide some information on past and ongoing processes, which can further serve as a basis for future predictions. The present work focuses on the comparison of permafrost soils with nearby non-permafrost soils in the Alpine (sites at 2700 m asl) and the sub-Alpine (sites at 1800 m asl) range of the Swiss Alps (Upper Engadine). We hypothesize that permafrost soils differ distinctly in their long-term soil erosion rates due to different water retention capacities.

The ^{10}Be abundance in a soil profile was estimated assuming that meteoric ^{10}Be is adsorbed in the fine earth fraction (<2 mm) and that ^{10}Be is deposited with precipitation. Soil erosion can be estimated by comparing the effective abundance of ^{10}Be measured in the soil with the theoretically necessary abundance for the expected age [1] using the equation

$$t_{corr} = -\frac{1}{\lambda}\ln\left(1-\lambda\frac{N}{q-\rho Em}\right)$$

m = measured concentration of ^{10}Be (atoms/g) in eroding horizons, ρ = bulk density (g/cm^3), t_{corr} = expected age, E = constant erosion rate, λ = decay constant of ^{10}Be (4.997 × 10^{-7}/a), N = inventory of ^{10}Be in the soil (atoms/cm^2) and q = annual deposition rate of ^{10}Be (atoms/cm^2/a).

The theoretical abundance of ^{10}Be was calculated for an expected soil age of 11 ka for the Alpine sites and 16 ka for the sub-Alpine sites, respectively [2] (Tab. 1).

Our results clearly show that soil erosion rates were higher in permafrost soils at both altitudes (Tab. 1; Fig. 1). Although Alpine soils have had a complex history and have been subjected to warmer and colder periods, differences in the thermal and hydrological conditions in soils strongly influenced erosional processes.

Sample	^{10}Be age [ka]	^{10}Be erosion rates [t/km^2/a]
PF$_{Alpine}$	5.3±0.3	35.9
PF$_{Subalpine}$	4.4±0.6	44.9
No PF$_{Alpine}$	11.6±1.5	~0
No PF$_{Subalpine}$	7.8±1.1	24.4

Tab. 1: *Estimated erosion rates from meteoric ^{10}Be for permafrost (PF) and non-permafrost (no PF) soils. ^{10}Be ages are calculated using equation*

$$t = -\frac{1}{\lambda}\ln\left(1-\lambda\frac{N}{q}\right)$$

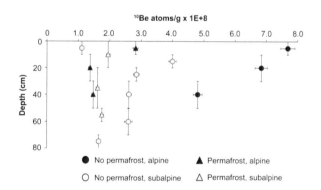

Fig. 1: *Depth profiles showing the accumulated meteoric ^{10}Be in the soil.*

[1] M. Egli et al., Geomorph. 119 (2010) 62
[2] J. Suter, Ph.D. thesis (1981) University of Zurich

[1] *Geography, University of Zurich*
[2] *Environmental Geosciences, University of Basel*
[3] *Geography and Geology, University of Würzburg, Germany*

^{10}Be IN SWISS LACUSTRINE SEDIMENTS

A record of solar activity?

M. Mann, J. Beer, F. Steinhilber, M. Christl

So far most attempts to reconstruct past solar activity were carried out on polar ice. We examined cosmogenic ^{10}Be in three varved lake sediment cores covering the last 100 years to investigate their suitability as a recorder of solar activity.

The ^{10}Be signal in lake sediments is composed of a component reflecting the radionuclide production in the atmosphere and a component related to the subsequent transport into the sediment. In order to separate these two components we applied singular-spectrum analysis (SSA).

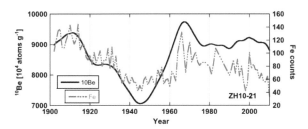

Fig. 1: *Transport modulated ^{10}Be signal in Lake Zurich sediments and the iron signal detected in this core (normalized XRF counts, annually resolved).*

Fig. 1 depicts the fraction of the measured ^{10}Be concentration which is assumed to represent the terrestrial (transport) modulation of the ^{10}Be signal in the sediment of Lake Zurich. This component shows very similar pattern to the XRF iron counts. In the production component of three different lacustrine sediments (Fig.2 a-c) we find a similar pattern as in the ice cores from NGrip [1] and Dye3 [2]. A cross-correlation analysis yields a significant negative correlation between the ^{10}Be production and the solar modulation potential. ^{10}Be lags the production on average by 1.5 years which corresponds to the expected transport time from the atmosphere to the Earth's surface. Hence, we

conclude that varved lake sediments are potentially suitable to study the solar activity of the past. However, the signal to noise ratio of the ^{10}Be production signal might be reduced by, e.g. geochemical processes (Fig. 1), and therefore, the interpretation of ^{10}Be data recovered from lacustrine sediments is not straightforward [3].

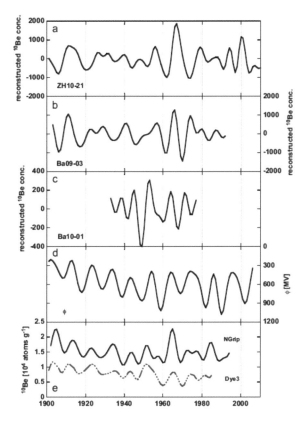

Fig. 2: *(a–c) Reconstructed ^{10}Be concentrations from Lake Baldegg and Lake Zurich. (d) The solar modulation potential. (e) The low-pass filtered NGrip [2] and Dye3 [1] ^{10}Be concentrations.*

[1] A.-M. Berggren et al., Geophys. Res. Lett. 36 (2009), L11801

[2] J. Beer et al., Nature 347 (1990), 164

[3] M. Mann et al., J. Atmosph. and Solar-Terrestr. Phys. 80 (2012) 92

^{10}Be AND THE LASCHAMP EVENT IN SEDIMENTS

^{10}Be analysis of detrital laminae of the Lisan Formation

R. Belmaker[1,2], M. Stein[2], J. Beer[3], M. Christl, D. Fink[4], B. Lazar[1]

^{10}Be ($t_{1/2}$=1.39 Ma) concentrations in sediments of lakes with large catchment areas depend not only on the production rate but also on climate related transport and erosion processes. In this study we evaluated the potential use of the annually laminated and accurately dated (by U-Th) lacustrine sediments of Lake Lisan (the late Pleistocene precursor of the Dead Sea) as a high resolution production rate archive of atmospheric ^{10}Be.

Lisan sediments comprise of annual pairs of primary aragonite and silty detritus material that originated from desert dust blown to the lake's vicinity and washed in with desert floods. ^{10}Be is mainly contained within detritus laminae.

The relative contributions of production and climate to the overall ^{10}Be signal were evaluated by measuring both the ^{10}Be and the chemical compositions of modern dust and of the detritus laminae (Fig. 1) in intervals representing lake level changes and in intervals representing rapid change in the ^{10}Be production (i.e. the Laschamp excursion).

Our results demonstrate that when the ^{10}Be production rate varies moderately, the ^{10}Be concentration correlates with silicate mineral proxies (R^2 = 0.84; Fig 1B) but not with the carbonate material (Fig. 1A). Yet during the Laschamp excursion (~41-40 ka BP), ^{10}Be concentrations show a about twofold increase (3.0±0.1 to 5.3±0.1×10^8 atoms·g^{-1}) that cannot be attributed to the mentioned correlations and possibly reflect the enhanced atmospheric production.

Radiocarbon measurements in aragonite of the Lisan Formation showed a significant Δ^{14}C anomaly across the Laschamp interval. We will combine ^{10}Be and ^{14}C data to deduce the non-production fraction of the radiocarbon anomaly.

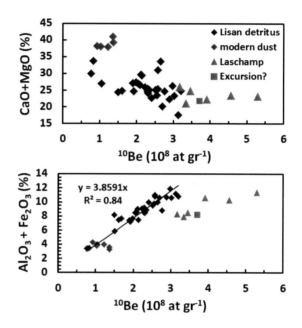

Fig 1: *Carbonate (A) and silicate (B) mineral proxies vs. ^{10}Be concentration. Samples that lie off the linear correlation line (red symbols) correlate in age with geomagnetic excursions (Fig 2).*

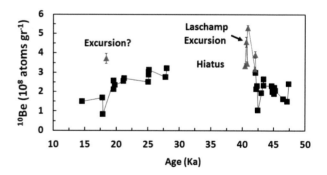

Fig. 2: *^{10}Be concentration vs. age*

[1] *Earth Sciences, Hebrew University, Jerusalem, Israel*
[2] *Geological Survey of Israel, Jerusalem, Israel*
[3] *Eawag, Dübendorf*
[4] *ANSTO, Menai, Australia*

"INSITU" COSMOGENIC NUCLIDES

In-situ ^{14}C production with depth

 When did the LGM ice mass decay in the high Alps?

The deglaciation history of the Gotthard Pass

Reconstructed Lateglacial ice extents in Valais

Late Pleistocene glacier advances in NE Anatolia

Exposure dating of the Chironico landslide

^{36}Cl exposure dating of "Marocche di Dro"

The AD 1717 rock avalanche near Mont Blanc

The age of the rock avalanche at Obernberg

Dating the Oeschinensee rock avalanche

Prehistoric rock avalanches at Rinderhorn

^{10}Be denudation rates in the Central Andes

Landscape response to tectonic uplift

Abated late Quaternary landscape downwearing

Landscape evolution of the Aare valley

Glacial impact on erosion of the Feldberg

IN-SITU ^{14}C PRODUCTION WITH DEPTH

A depth profile of in-situ produced ^{14}C along a quartzite core

M. Lupker[1], K. Hippe[1], F. Kober[1], L. Wacker, R. Braucher[2], D. Bourlès[2], J. Vidal Romani[3], R. Wieler[1]

In-situ produced ^{14}C has recently emerged as a complement to other longer-lived cosmogenic nuclides such as ^{10}Be or ^{26}Al. The relatively short half-life (5730 a) of ^{14}C makes it suitable to investigate surface processes such as denudation rates or sediment residence times on ka scales. The wide application of in-situ ^{14}C for quantitative studies is however bound to the proper calibration of its production mechanisms and rates. As other cosmogenic nuclides, ^{14}C is produced at the Earth's surface by nuclear reactions with incoming neutrons and muons.

The production rate of ^{14}C has been determined for quartz exposed at the surface where neutrons dominate the overall production [1]. At depth, however, the muon production pathway starts to dominate, because the mean attenuation length of muons is considerably longer than that of neutrons. So far, the muon derived in-situ ^{14}C production rate is solely based on theoretical and experimental work [2] that has not been tested on natural objects.

We measured the ^{14}C concentration in quartz along the Leymon High core (northwestern Spain) using the ETH ^{14}C extraction line [3] and the MICADAS (gas ion source) AMS facilty.

This core has been drilled down to 20 m in a quartz dyke and has already been used to refine the depth-dependent production rate of ^{10}Be and ^{26}Al [4]. Our results on 14 samples of this core spanning a depth range from 1 to 1545 cm allow us to estimate the muogenic contribution to the overall ^{14}C concentrations measured along the core (Fig. 1). This data set yields a local surface muon production rate slightly lower than the previous estimates [2]. Further measurements on other cores should be carried out to verify these estimates.

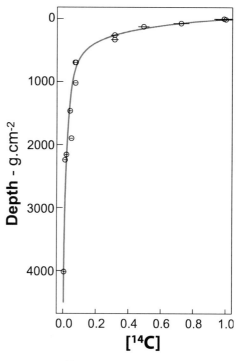

Fig. 1: *In-situ ^{14}C concentration, normalized to the surface concentration, in quartz samples from the Leymon High core.*

[1] N.A. Lifton et al., Geochim. Cosmochim. Acta 65 (12) (2001) 1953

[2] B. Heisinger et al., Earth Planet. Sci. Lett. 200 (2002) 357

[3] K. Hippe et al., Nucl. Instr. & Meth. B 294 (2013) 81

[4] R. Braucher et al., Nucl. Instr. & Meth. B 294 (2013) 484

[1] *Earth Sciences, ETH Zurich*
[2] *CEREGE, Aix-en-Provence, France*
[3] *Geology, University of La Coruña, Spain*

WHEN DID THE LGM ICE MASS DECAY IN THE HIGH ALPS?

[10]Be surface exposure dating of the LGM ice surface lowering

C. Wirsig, S. Ivy-Ochs, N. Akcar[1], J. Zasadini[2], M. Christl, C. Schlüchter[1]

The extent of ice in the Alps during the Last Glacial Maximum (LGM) is well known based on extensive mapping efforts over the previous century. In Switzerland, massive ice domes in the high Alps fed glaciers that advanced far into the foreland [1]. Constraining the timing of events proves to be a more challenging task. [10]Be surface exposure ages and radiocarbon ages imply that the piedmont glaciers in the Swiss Alpine foreland had retreated before 19 ka BP [2]. This result is in excellent agreement with the end of the last glacial period spanning from 30–19 ka BP, as inferred from stable isotopes in marine and polar ice cores. Did the ice surface in the high Alps decay simultaneously?

Cosmogenic nuclide dating determines the time that a surface has been exposed, in this case since its previous cover of ice vanished. It can thus date this event directly, in contrast to radiocarbon, OSL or other dating methods. We have collected samples for [10]Be analyses from several sites in the eastern, central and western Alps to address the title's question.

Fig. 1: *Study site Gelmersee in Oberhaslital: samples GELM1-5 were taken from the exposed ridge at the northern shore (©googleEarth).*

So far, the small number of studies using the same methodology report ages that are no older than 17.3 ka [3]. The age discrepancy compared to the ice retreat from the foreland and to the results from marine and polar ice cores is assumed to be caused by the influence of temporary sediment or ice cover on the samples.

Fig. 2: *Sample GELM3: 3m tall erratic boulder.*

Considering the circumstances encountered by previous studies, we took particular care choosing sample sites, such as exposed ridges, to avoid ambiguities in interpretation. In Oberhaslital, for example, we sampled bedrock outcrops and boulders just below the trim line on the steep north-eastern mountain sides above Gelmersee (Fig. 1). Sampling sites such as GELM3 (Fig. 2) were selected to avoid places of possible accumulation of local ice or sediment and to exclude the influence of cirque glaciers flowing into the main valley.

[1] D. Florineth and C. Schlüchter, Eclogae Geologicae Helvetiae 91 (1998) 295
[2] S. Ivy-Ochs et al., Eclogae Geologicae Helvetiae 97 (2004) 47
[3] K. Hippe, PhD thesis (2012) ETH Zürich

[1] *Geology, University of Bern*
[2] *Geology, AGH University, Krakow, Poland*

THE DEGLACIATION HISTORY OF THE GOTTHARD PASS

Surface exposure dating by combined [10]Be and *in situ* [14]C

K. Hippe[1], S. Ivy-Ochs, F. Kober[1], J. Zasadni[2], R. Wieler[1], L. Wacker, P.W. Kubik, C. Schlüchter[3]

Rapid downwasting of the large piedmont glaciers in the Alpine foreland marks the onset of the Alpine Lateglacial and the beginning of gradual climate warming after the Last Glacial Maximum (LGM). To improve the understanding of the effect of such climate change on the high Alpine mountain glaciers, we have performed surface exposure dating using cosmogenic [10]Be and *in situ* [14]C analysis of bedrock surfaces on the Central Alpine Gotthard Pass, Switzerland. Dating was combined with detailed mapping of glacial erosional features (Fig. 1). These give evidence for gradual glacier downwasting from the maximum LGM ice volume causing a progressive re-organization of the paleoflow pattern and a southward migration of the ice divide.

evident from the erosional features and is consistent with another group of [10]Be exposure ages of 12-13 ka. These are correlated with the decay of comparatively small glaciers of the Younger Dryas Egesen stadial. Dating of a boulder close to pass elevation gives a minimum age of 11.1 ka for final deglaciation by the beginning of the Holocene. *In situ* [14]C data are overall in good agreement with the [10]Be ages. This consistency excludes the presence of significant glacial ice on Gotthard Pass during the Holocene and points to a continuous exposure since the end of the Younger Dryas. From the combination of both nuclides, we were able to evaluate the necessity of a snow shielding correction and gained constraints on the amount of Holocene snow cover (Fig. 2).

Fig. 1: *Crescentic gouges and glacial polish in the Gotthard Pass area. Arrow gives paleoflow direction.*

The oldest exposure ages obtained by [10]Be (16-15 ka; snow corrected) are interpreted to reflect the decay of the large Gschnitz glacier system and post-date deglaciation of the foreland by a few thousand years. Thus, continuous ice cover and glacier transfluence over the pass may have persisted throughout the Oldest Dryas, possibly until the onset of the Bølling warming. A younger phase of local glacier re-advance is

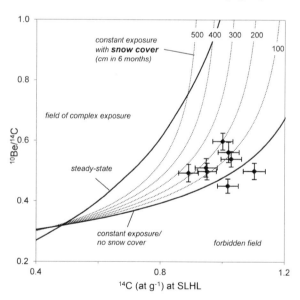

Fig. 2: *Combined [10]Be-^{14}C data in a two-nuclide diagram indicating constant exposure with snow cover for most sampling sites.*

[1] *Earth Sciences, ETH Zurich*
[2] *Geology, AGH University, Kraków, Poland*
[3] *Geological Sciences, University of Bern*

RECONSTRUCTED LATEGLACIAL ICE EXTENTS IN VALAIS

^{10}Be dating supports numerical modeling of Lateglacial ice advances

K. Leith[1], J. Moore[1], P. Sternai[1], S. Ivy-Ochs, F. Hermann[1], S. Loew[1]

Remnant moraine deposits preserved on the walls of inner Alpine valleys provide valuable constraint on the timing and extent of Alpine glaciers during Lateglacial stadia. The Mattertal and Saastal (Canton Valais, Switzerland) were major tributaries to the Rhone Glacier during the Last Glacial Maximum. They are oriented approximately north-south, and sub-parallel, 10 km apart. The total catchment area of the Mattertal is 450 km^2, while the Saastal has a catchment area of 250 km^2. Valley floor elevations in each vary between 700 m and 2200 m. Detailed mapping of a prominent moraine sequence within the region has allowed us to delineate an extensive Lateglacial re-advance in each valley that coincides with a clear, consistent change in the geomorphology of both. Moraine deposits interpreted to reflect the terminus of the principal valley glaciers are located at 1300 m elevation in the Mattertal, and 1650 m in the Saastal, with lateral moraine deposits located between 300 m and 500 m above the present-day valley floor.

Fig. 1: *Boulder dated to constrain the depositional age of a near-terminal moraine in the Saas Valley.*

Although the geography of the two valleys is similar, the mapped extents reflect complicated dendritic glacier systems, and calculations of equilibrium line altitude enabling correlation of the stadia are difficult. Cosmogenic ^{10}Be exposure age dating is used to constrain the timing of moraine emplacement to the Egesen stadia (Fig. 1).

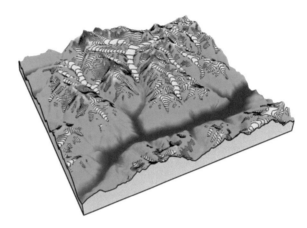

Fig. 2: *Modeled Egesen I glacier extent calibrated from ^{10}Be exposure ages.*

We use the numerical landscape evolution model ICE-CASCADE [1] to model ice extents for an equilibrium line altitude (ELA) characteristic of this stadial, finding a close correlation between modeled and mapped extents in the two valleys (Fig. 2). By reducing model temperatures, we can then use the calibrated model to compare mapped moraine deposits to model ELA depressions, matching the results to literature values for older Lateglacial stadia. This provides important insights into the Lateglacial history of the region, as well as the dynamics of ice advance during similar interglacial - glacial transitions in the valleys.

[1]   J. Braun et al., Ann. Glaciol. 28 (1999) 282

[1] *Geology, ETH Zurich*

LATE PLEISTOCENE GLACIER ADVANCES IN NE ANATOLIA

Surface exposure dating with cosmogenic ^{10}Be and ^{26}Al

R Reber[1], D. Tikhomirov[1], N. Akçar[1], S. Yeşilyurt[2], V. Yavuz[3], P.W. Kubik, C. Schlüchter[1]

In order to extend our knowledge on the chronology of late Pleistocene glacier advances in Anatolia, we sampled 40 erratic boulders for surface exposure dating with in-situ produced cosmogenic ^{10}Be and ^{26}Al in the Basyayla valley, a small tributary valley in north eastern Turkey (Fig. 1). This typical U-shaped glacial valley contains moraines which suggest five distinct glacier advances.

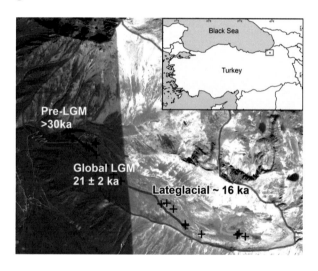

Fig. 1: *Satellite image of the Basyayla valley, locations of sampled boulders marked by crosses.*

^{10}Be exposure ages show at least one advance at around 35 ka, which extended down to an altitude of 2350 m a.s.l. A subsequent advance occurred at around 21 ka and reached to 2480 m a.s.l. (Fig. 2). The Lateglacial advance at around 16 ka was limited to the environs of a cirque glacier at an altitude of 3050 m a.s.l. Since then, the valley was free of ice. ^{10}Be and ^{26}Al concentrations from the bedrock sample indicate a simple exposure history and that there was no ice in the adjacent valley at least during the last 37 ka.

In our study, we exposure dated, for the first time, a well constrained LGM (Last Glacial Maximum; 21±2 ka) terminal moraine in NE Anatolia (Fig. 2). In addition, a larger glacial extent has been exposure dated to > 30 ka, prior to the global LGM.

Fig. 2: *Terminal moraine of "Global LGM" view up valley.*

Our results correlate well with data from the neighboring Çoruh, Kavron [1] and Verçenik [2] valleys and from the Uludağ [3], as well as with the existing chronologies from other Anatolian Mountains [4]. Our study provides a better understanding of glaciations in Anatolia prior to and during the global LGM at 21±2 ka, as well as during the Lateglacial.

[1] N. Akçar et al., Quat. Int. 164-165 (2007) 170
[2] N. Akçar et al., J. Quat. Sci. 23/3 (2008) 273
[3] C. Zahno et al., Quat. Sci. Rev. 29 (2010) 1173
[4] Sarıkaya et al., Dev. in Quat. Sci. 15 (2011) 393

[1] *Geology, University of Bern*
[2] *Geography, Çankırı University, Turkey*
[3] *Geological Engineering, Istanbul Technical University, Turkey*

EXPOSURE DATING OF THE CHIRONICO LANDSLIDE

A Bølling-Allerød crystalline landslide in the Alps

A. *Claude[1], S. Ivy-Ochs, F. Kober, M. Antognini[2], B. Salcher, P.W. Kubik*

The abundance of slope instabilities in the European Alps during the Lateglacial and Holocene, suggests marked postglacial landscape response. In this study, we investigated the 530 million m^3 large Chironico landslide in the Leventina valley in the southern Swiss Alps. The sliding mass, consisting of Leventina granitic gneiss, detached from the eastern valley wall. It was deposited right at the Ticinetto stream mouth, creating a northern and southern lobe, as well as damming the Ticino River. Wood fragments originating from basal lacustrine sediments in an upstream-dammed lake yielded a minimum age for the landslide of around 13,500 cal yr BP [1].

In order to directly date the landslide, we sampled 14 suitable boulders (Fig. 1) for surface exposure dating with the cosmogenic nuclides ^{10}Be and ^{36}Cl.

Fig. 1: *Illustration of four sampled boulders.*

The ^{10}Be ages were calculated with the NE North America developmental version of the CRONUS-Earth online calculator [2]. For the ^{36}Cl sample, the in-house developed Matlab age calculator was used [3]. The results revealed that the surface exposure ages are similar to the radiocarbon ages. Moreover they showed that the northern and southern depositional lobes result from one single event. We replicated the runout distance and lateral expansion of the landslide deposit by applying a numerical model using the program DAN 3D (Fig. 2).

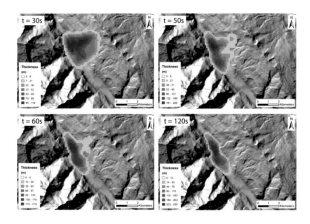

Fig. 2: *Result of the landslide runout modeling. The moving mass is shown at four different time steps.*

The runout modeling further confirmed the location of the detachment scarp on the eastern valley flank. The absolute age determination revealed that the Chironico landslide was released during the Bølling-Allerød interstadial, making it one of the oldest dated instabilities in the Alps.

[1]   M. Antognini and R. Volpers, Bull. Appl. Geol. 7(2) (2002) 113

[2]   G. Balco et al., Quat. Geochr. 4 (2009) 93

[3]   V. Alfimov and S. Ivy-Ochs, Quat. Geochron. 4 (2009) 462

[3]   O. Hungr, Can. Geotech. J. 32 (1995) 610

[1] *Geology, University of Bern*
[2] *Museo cantonale di storia naturale, Lugano*

^{36}Cl EXPOSURE DATING OF "MAROCCHE DI DRO"

A polyphase rock avalanche in the Sarca Valley (Trentino, Italy)

S. Ivy-Ochs, S. Martin[1], P. Campedel[2], V. Alfimov, C. Vockenhuber, A Viganò[3], G. Carugati[4], A. Fontanari[1]

Numerous rock avalanche deposits of undetermined age are located along the Adige and Sarca valleys of the Eastern Alps (Trentino, Italy) [1 and references therein]. The *Marocche di Dro* is one of the most extended of these deposits, with an area of about 13 km^2, a volume of about 10^9 m^3, and a runout of more than 8 km. These rock avalanche deposits cover the middle Sarca Valley alluvial sediments, north of Lake Garda (Fig. 1).

Fig. 1: View of the middle Sarca valley with the "Marocche di Dro" rock avalanche deposits and postulated main scarps of Mt. Brento. View is to the NNE.

The *Marocche di Dro* comprises deposits of three large rock avalanches: the *Lago Solo* body to the south, the *Marocca di Kas* in the middle and the *Marocca principale* in the northern sector. The deposits are differentiated based on relative stratigraphy, type of vegetal coverage, and degree of karst development on the boulders. Postulated main scarps are located on the western side of the Sarca Valley, along the steep faces of Mt. Brento and Mt. Casale. The presence of these scarps is strictly related to the Southern Giudicarie fault system, that here is constituted by regular NNE-directed ESE-vergent thrust faults.

Fig. 2: Sampled boulder of the "Marocche di Dro" rock avalanche deposits. Note the chert stringers on the face of the limestone block, which may allow us to analyze for ^{10}Be as well ^{36}Cl.

The rock avalanche developed within carbonate rocks of Mesozoic age, mainly (dolomitic) limestones of the Jurassic *Calcari Grigi* Group (Fig. 2). Preliminary ^{36}Cl exposure dating results for boulders of the middle and northern deposits suggests middle and late Holocene ages. The latter are in part comparable with post-Roman time ages proposed by Trener [2] based on the presence of Roman relics at the base of the youngest deposit.

[1] S. Martin et al. Quat. Geochron., subm.
[2] G. B. Trener, Geologia delle Marocche, Studi Trent. Sci. Nat. 34 (1924) 319

[1] *Geoscience, University of Padua, Italy*
[2] *Geological Survey of the Province of Trento, Trento, Italy*
[3] *National Institute of Oceanography and Experimental Geophysics, CRS, Udine, Italy*
[4] *Chemical and Environmental Sciences, University of Insubria, Como, Italy*

THE AD 1717 ROCK AVALANCHE NEAR MONT BLANC

Dilemma of distinguishing rock avalanche and glacial deposits

N. Akçar[1], P. Deline[2], S. Ivy-Ochs, V. Alfimov, I. Hajdas, P.W. Kubik, M. Christl, C. Schlüchter[1]

The extent of the rock avalanche deposits in the upper Ferret Valley (Mont Blanc Massif, Italy; Figs. 1 and 2), which occurred on September 12th, AD 1717, has been under debate for a long time. Although this rock avalanche was historically recorded, detailed maps were not made at the time. Later investigators attributed the accumulation of granitic boulders and irregular ridges covering the upper valley floor to either deposition by a Lateglacial glacier, to the AD 1717 rock avalanche, or to a complex mixture of glacial deposition, earlier rock avalanche and the AD 1717 rock avalanche [1].

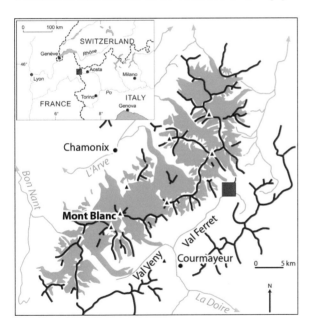

Fig. 1: *Sketch map of the Mont Blanc Massif. Blue color shows the glaciers and red square the study area.*

We applied cosmogenic [10]Be to end this dilemma; we sampled 16 boulders of this deposit and three from boulders outside of this deposit for surface exposure dating. Exposure ages of the granitic boulders within the upper valley deposit vary between 300 and 500 years within the limits of error. Our results show that at least 14 of these boulders were deposited by the AD 1717 rock avalanche, and its deposits do cover the whole upper Ferret valley floor (Fig. 2). These correlate well with the results from dendrochronology, lichenometry and existing radiocarbon ages from wood samples, however not with the older ^{14}C results from peat bog in Plan d'Arp-Nouvaz in the upper part of the valley (Fig. 2).

Fig. 2: *The upper Ferret Valley. View towards the southwest.*

With this study, we directly contribute to distinguish between rock avalanche deposits and older moraine sets in valleys of formerly glaciated mountains, two assemblages of landforms which often look alike, as well as to the assessment of the natural risks in the Mont Blanc Massif area.

[1]   P. Deline and M. Kirkbride, Geomorph. 103 (2009) 80

[2]   N. Akçar et al., J. Quat. Sci. 27 (2012) 383

[1] *Geology, University of Bern*

[2] *EDYTM Laboratory, University of Savoie, Chambery, France*

THE AGE OF THE ROCK AVALANCHE AT OBERNBERG

Surface exposure dating of boulders with ^{36}Cl

M. Ostermann[1], D. Sanders[1], S. Ivy-Ochs, V. Alfimov, M. Rockenschaub[2], A. Römer[2]

In the Obernberg valley (Tyrol, Austria) the character of a rock avalanche deposit led to diverse interpretations for more than a hundred years. Based solely on the morphology of ridges and hillocks, the landforms were thought to be terminal moraines and kames (Figs. 1 and 2). We applied surface exposure dating and radiocarbon dating to determine the age of the catastrophic rock slope failure [1]. The rock avalanche involved an initial rock volume of about 45 million m³, with a runout of 7.2 km over a total vertical distance of 1330 m (Fahrböschung 10°).

Fig. 1: *Ridges and hillocks at the distal part (~ 2 km) of the rock avalanche mass that formerly were interpreted as moraines until we determined ^{36}Cl exposure ages.*

A minimum-age constraint of the mass-wasting event was obtained by radiocarbon dating of organic remains found in alluvial fan deposits on top of the rock avalanche deposits (VERA-4980 6980±45 ^{14}C yr BP; 7931 to 7699 cal yr BP) [1].

From ^{36}Cl surface exposure dating of four boulder surfaces we obtained the following ages: 9.16±0.40 ka (OB1), 12.09±0.55 ka (OB2), 8.24±0.60 ka (OB3), and 8.32±0.40 ka (OB4). Ages were calculated using the production rates

listed in Alfimov and Ivy-Ochs [2]. We attribute the outlier age, 12.09±0.55 ka, to inheritance which is often observed in rock avalanche boulders. The average age of the remaining three ages, 8.6±0.6 ka, indicates an early Holocene age for the rock avalanche event. The purported glacial landforms actually accumulated from the rock avalanche.

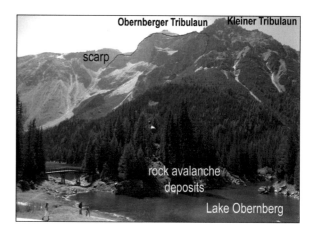

Fig. 2: *Scarp area and upper accumulation area of the Obernberg rock avalanche.*

The transversal ridges are arranged into two highly regular higher-order waves, each of which consists of waxing and shrinking ridges. We suggest that the arrayed ridges reflect a mechanical aspect of the movement, perhaps propagation of waves towards the snout of the avalanche deposit.

[1] M. Ostermann et al., Geomorph. 171-172 (2012) 83

[2] V. Alfimov and S. Ivy-Ochs, Quat. Geochron. (2009) 462

[1] *Geology, University of Innsbruck, Austria*
[2] *Geological Survey of Austria, Vienna, Austria*

DATING THE OESCHINENSEE ROCK AVALANCHE

^{36}Cl exposure dating of rock avalanche deposits

P. Köpfli[1], J.R. Moore[1], S. Ivy-Ochs, C. Vockenhuber, S. Knapp[1], A. Gilli[1], I. Hajdas

Some of the most dramatic examples of Alpine rock avalanches can be found in the Kandersteg area of the central Swiss Alps. Most of these failures have been previously interpreted to be Lateglacial or early Holocene in age, however the precise failure timing is rarely known. We investigate the case study of the Oeschinensee rock avalanche (Fig. 1), whose deposit dams the Lake Oeschinen in the UNESCO world heritage site Jungfrau-Aletsch-Bietschhorn.

The Oeschinensee rock avalanche deposit, and formation of Lake Oeschinen, was previously thought to be of early Holocene age, however the present study shows that this is not the case. Cosmogenic ^{36}Cl exposure ages from seven boulders, combined with radiocarbon dates from wood samples on top of the rock avalanche deposits in the lake, reveal a much younger age near the beginning of the Subatlantic, which is approaching the period of historical record in the area.

Fig. 1: *The Oeschinensee rock avalanche with source area (blue) and partial view of deposit (red) indicated. Debris dams the Lake Oeschinen.*

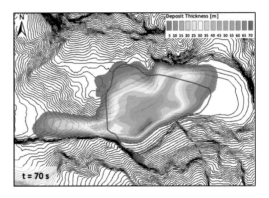

Fig. 2: *Deposit geometry and thickness near the end of simulated runout – red line is the mapped deposit extent, dashed blue is the release area.*

Combined field investigations and lake sediment coring [1] revealed a deposit volume of roughly 50 million m^3, which dammed the lake and is more than 60 m thick in places. The release area consists of a steeply inclined dip-slope sliding plane in massive limestone on the southern side of the valley. Field reconstruction and runout simulation using a quasi-3D continuum dynamic analysis code show that the rock avalanche first travelled across the valley, ran up the opposite slope, and then spread in both downstream and upstream directions (Fig. 2). Interactions with older mass movement deposits were crucial in controlling the runout dynamics and final deposition pattern.

The large lag time between the neighboring Kandertal rock avalanche indicates that the two events did not follow in quick succession as previously suspected. The Oeschinensee rock avalanche was also not the last large rock slope failure that occurred in the valley; there are at least four other clearly visible release areas with deposits traceable to the lake bed, all with a volume larger than 200'000 m^3 and all of which occurred in the last few thousand years.

[1] S. Knapp et al., Laboratory of Ion Beam Physics Annual Report (2012) 38

[1] *Earth Sciences, ETH Zurich*

PREHISTORIC ROCK AVALANCHES AT RINDERHORN

Surface exposure dating and runout modeling of rock avalanches

L. Grämiger[1], J.R. Moore[1], C. Vockenhuber, S. Ivy-Ochs

Large prehistoric rock avalanches are frequently associated with the retreat of Alpine glaciers following the Last Glacial Maximum (LGM). However, due to a lack of dated rock avalanche deposits in the Alps (cf. [1]), conclusions regarding the temporal occurrence of these events remain elusive. Here are presented two case studies of rock avalanches in the Rinderhorn area of the central Bernese Alps, Switzerland.

Fig. 1: *Release and parts of the deposit area of the Klein Rinderhorn rock avalanche.*

Fig. 2: *Runup hill at Schwarenbach.*

The Klein Rinderhorn rock avalanche (Fig. 1 and 2) released approximately 52 million m^3 of sedim-entary rock with a maximum runout distance of 4.3 km. Morphological observations and radio-carbon dating of sediments on top of the rock avalanche deposit suggest that the slope failed during the Holocene between the Preboreal and the end of the post-glacial climatic optimum. Cosmogenic ^{36}Cl surface exposure dating of boulders in the rock avalanche deposits indicates a Boreal age and confirms the hypothesis of failure during Holocene climate warming.

The 4.6 million m^3 Daubensee rock avalanche (Fig. 3) has a maximum runout distance of 2.5 km, breaching a Lateglacial lateral moraine. Concluding from runout modeling results, the Daubensee rock avalanche most likely failed at the end of the Egesen Lateglacial stadial, when the snout of the Lämmeren glacier was located near the front of the Daubensee. Cosmogenic ^{36}Cl surface exposure dating indicates a Preboreal age and therefore supports this hypothesis.

Fig. 3: *Release area of the Daubensee rock avalanche below the Rinderhorn summit.*

The lag time between deglaciation and failure suggests that both events were likely not directly triggered by deglaciation at the end of the LGM. Glacial erosion certainly conditioned the failures by exposing steep slopes in NW-dipping limestone. However, progressive failure during Holocene climate warming appears to be the predominant factor leading to the Rinderhorn rock avalanches.

[1] C. Prager et al., Austrian J. Earth Sci. 102 (2009) 4.

[1] *Geology, ETH Zurich*

^{10}Be DENUDATION RATES IN THE CENTRAL ANDES

Control of climate and tectonic on medium-term denudation rates

F. Kober[1], G. Zeilinger[2], K. Hippe, R. Grischott[1], T. Lendzioch[2], M. Christl

The Central Andes are an ideal place to study the effects of climate and/or tectonics on short- and long-term denudation rates. There, the impact on surface processes of the recent change towards wetter climate at the end of the Pleistocene and the eastward migrating deformation front can be gauged.

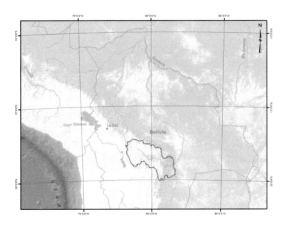

Fig. 1: *Rio Grande catchment central Bolivia*

We determined cosmogenic ^{10}Be catchment-wide denudation rates on 58 active channel sediment samples at sites in the Rio Grande catchment in the Central Andes of Bolivia (Fig. 1). The sub-catchments traverse from west to east the Eastern Cordillera (EC), the Inter-Andean Zone (IAZ) and the Sub-Andes (SA). Catchment-wide denudation rates within the Rio Grande catchments range from 7 mm/ka to 1550 mm/ka (Fig. 2), with a mean of 260 mm/ka. The higher denudation rates are observed in the humid SA in accordance with the active deformation front but also within the locus of the marked Holocene climate change. In the arid to semiarid EC and IAZ, we obtained mean rates of 88 and 60 mm/ka, respectively. Sediment budget analysis suggests that sediment mixing with respect to the cosmogenic nuclide method may not be perfect. The ^{10}Be-based rates also contrast with earlier

established sediment yield data. Interestingly, high sub-catchment rates of the SA do not appear to have an effect on the rates determined further downstream in an area-weighted mode.

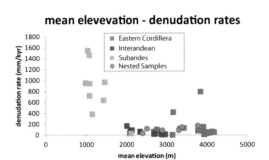

Fig. 2: *^{10}Be denudation rates for sub-catchments*

A clear correlation of the significantly higher mean denudation rates of 850 mm/ka in the SA with geomorphic or climatic parameters is not readily apparent. However, recent activity along local fault zones in the SA may cause enhanced sediment fluxes. This fault activity pattern continues to the present day as shown by active deformation as reflected in recent shallow seismicity clusters and enhanced shortening rates revealed by GPS measurements.

The absence of correlation of geomorphic parameters with our denudation rates is potentially caused by the size of the sampled catchments where multiple surface processes prevent clear morphometric signals. The influence of active deformation on geomorphic parameters in the Bolivian Andes is the subject of further studies.

---

[1] *Geology, ETH Zurich*
[2] *Geology, University of Potsdam, Germany*

LANDSCAPE RESPONSE TO TECTONIC UPLIFT

Insights on geomorphic response from low uplift rate regimes

V. Vanacker[1], N. Sougnez[1], P.W. Kubik

Geomorphic response to tectonic uplift has been studied extensively for tectonic settings with high uplift rates. In contrast, few data exist for tectonic settings with low uplift rates. This research focuses on the relation between tectonic activity, river and slope morphology, and denudation rates for the Belgian Ardennes Massif (Fig. 1).

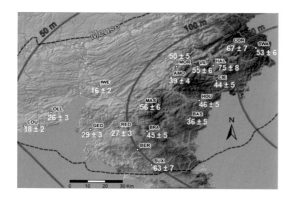

Fig. 1: *Spatial pattern of ^{10}Be-derived denudation rates (mm/ka) for the Ardennes Massif (Belgium). The tectonic uplift isolines (red lines, m) are constructed based on the terrace stratigraphy of the Meuse River [1].*

The east-west variation in uplift rates in this Paleozoic massif affords a good opportunity to unravel several key questions regarding extrinsic controls on landscape evolution. We selected 20 3rd-order catchments that cover various tectonic domains with uplift rates ranging from $\approx$ 0.06 to 0.20 mm/ka since mid-Pleistocene times [1].

^{10}Be-derived denudation rates vary between ca. 15 and 80 mm/ka, and an east-west gradient in long-term denudation rates can be observed. Although we observe that the high-uplift region has generally higher ^{10}Be denudation rates, an important variability in denudation rates exists.

Morphometric analyses of river and slope morphology indicate that the topography of 3rd-order catchments is not yet in topographic

steady state, as it exhibits clear knick zones in slope and river profiles. The spatial variation that we observe in slope and channel morphology between the 20 3rd-order catchments is not only the direct result of the differential uplift pattern, but also reflects the transient adjustment of river channels to tectonic forcing (Fig. 2).

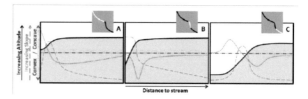

Fig. 2: *Schematic representation of change in slope morphology measured perpendicular to the river channel for three different positions: (A) upstream of the channel knick zone, (B) in the knick zone and (C) downstream of the knick zone.*

The Meuse River that is draining the third-order catchments acts as the local base level to which the fluvial system is adapting after the post 0.73 Ma of tectonic uplift. Our morphometric analysis suggests that the response of the fluvial system was strongly diachronic, and that a transient signal of adjustment is migrating from the Meuse valley towards the Ardennian headwaters. This hypothesis is consistent with the recent insights from this ^{10}Be-derived denudation rate study.

[1]   A. Demoulin and E. Hallot, Tectonophysics. 474 (2009) 696

[2]   N. Sougnez and V. Vanacker, HESS. 15 (2011) 1095

---

[1] *Earth and Life Institute, Université de Louvain, Belgium*

ABATED LATE QUATERNARY LANDSCAPE DOWNWEARING

Postglacial denudation outpaced by long-term exhumation, NW India

H. Munack[1], J. Blöthe[1], D. Scherler[1], H. Wittmann[2], P.W. Kubik, O. Korup[1]

In the NW Himalayas of India and Pakistan, areas of very rapid exhumation and bedrock river incision, such as the Indus gorge of the western Himalayan syntaxis, are juxtaposed to areas of extremely low denudation of the western Tibetan Plateau (Fig. 1). The Indus River, one of Asia's premier rivers, links the low-relief Tibetan Plateau areas to the steep and rugged Trans-Himalayan ranges draining along the Indus-Tsangpo suture zone. There, the Ladakh Batholith, testimony to an ancient island arc system accreted to the Eurasian margin, and the Zanskar Range, built up by Indus Group fore-arc basin sedimentary rocks, are the prominent lithologies that flank the course of the river.

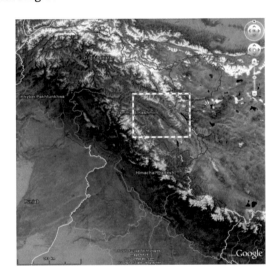

Fig. 1: *Location of the upper Indus River valley between the NW Himalaya and the western Tibetan Plateau (Google Earth).*

Data on denudation rates in this arid bedrock landscape are sparse. Here we report on a regional inventory of basin-wide ^{10}Be derived denudation rates from 33 tributaries distributed along ~300 km of the upper Indus River. We find that denudation rates range between 10 and 110 mm ka^{-1}, decreasing upstream by an order of magnitude along this course (Fig. 2).

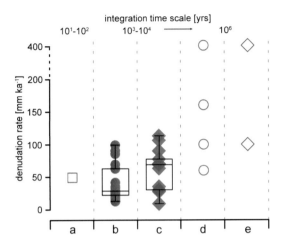

Fig. 2: *Denudation data from the upper Indus region. Red: Ladakh Batholith; Blue: Indus group; a) historical data [1]; b & c) cosmogenic ^{10}Be derived data (this study), whiskers to 5th/95th percentile; d & e) long-term denudation estimates from thermochronometer data [2-4].*

Our data integrate over $10^3 - 10^4$ years, and can be considered postglacial [5]. Compared to long-term estimates from thermochronometers, they are systematically lower on average. We infer that denudation rates along the western Tibetan Plateau margin must have been higher earlier in the Pleistocene, and decayed since the late Quaternary.

[1] E. Garzanti et al., Earth Planet. Sci. Lett. 229 (2004) 287

[2] H. Sinclair and N. Jaffey, J. Geol. Soc. London 158 (2001) 151

[3] R. Kumar et al., Curr. Sci. 25 (4) (2007) 490

[4] L. Kirstein, Tectonophysics 503 (2011) 222

[5] L. Owen et al., GSA Bull. 118 (2006) 383

[1] *Earth and Environmental Sciences, University of Potsdam, Germany*
[2] *Helmholtz-Zentrum Potsdam, Deutsches GeoForschungsZentrum GFZ, Potsdam, Germany*

LANDSCAPE EVOLUTION OF THE AARE VALLEY

[10]Be-derived erosion rates from catchments in the Bernese Alps

C. Glotzbach[1], C. Wangenheim[1], A. Hampel[1], P.W. Kubik

The present-day topography of the Alps is the result of the complex interplay between tectonic uplift, climate change and erosional processes. Repeated Quaternary glaciations caused by climate variations result in increased rates of valley incision and an overall increase in local relief [1, 2, 3].

In this project we studied the impact of major glaciations on landscape evolution applying geomorphological analysis of catchments in the Bernese Alps for which we determined catchment-wide denudation rates with cosmogenic [10]Be (Fig. 1).

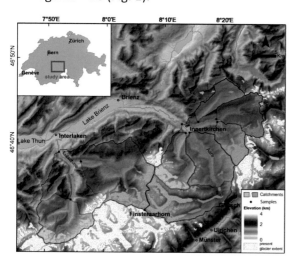

Fig. 1: *Topographic map of the study area with sample locations/catchments.*

Longitudinal channel profiles differ strongly from fluvial steady-state concave profiles, but show distinct steps with increased steepness indices typical for glacial impacted profiles [4]. Steps are partly controlled by lithology and show a strong correlation with former glacier positions (Fig. 2).

[10]Be-derived catchment-wide erosion rates range from ~80 to ~1400 mm/ka with most rates around 500 mm/ka. A sample from the

Aare river upstream of Innertkirchen yields the highest erosion rate, which is obviously influenced by frequent debris-flows from oversteepened hillslopes [5]. The geomorphic analysis reveals that erosion rates generally increase with morphometric measures (slope, steepness index).

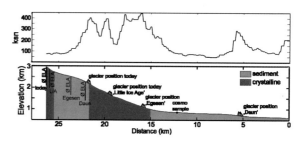

Fig. 2: *Longitudinal channel profile with steepness indices (ksn), glacier position and equilibrium line altitude (ELA) of a sample taken in the Lütschine south of Interlaken.*

Based on this data we suggest that [10]Be-derived erosion rates in the study area reflect the transient erosional response to glacially formed geomorphological inheritance.

[1] P. Häuselmann et al., Geology 35 (2007) 143.

[2] C. Glotzbach et al., Earth Planet. Sci. Lett. 304 (2011) 417

[3] P.G. Valla et al., Nature Geosci. 4 (2011) 688

[4] K.X. Whipple et al., Nature 401 (1999) 39

[5] F. Kober et al., Geology 40 (2012) 935

[1] *Geology, University of Hannover, Germany*

GLACIAL IMPACT ON EROSION OF THE FELDBERG

[10]Be catchment-wide erosion rates from the southern Black Forest

C. Glotzbach[1], R. Röttjer[1], A. Hampel[1], P.W. Kubik

The southern part of the Black Forest mountain region in southwest Germany was glaciated during the Last Glacial Maximum (LGM). Alpine/valley glaciers carved deep morphologically distinct valleys with 'oversteepened' hillslopes around the Feldberg region (Fig. 1).

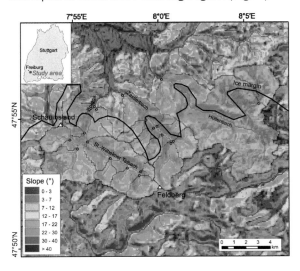

Fig. 1: *Slope map of the Feldberg region with sample locations/catchments and LGM ice margin.*

In this study we determined catchment-wide [10]Be-derived erosion rates of stream sediments of formerly glaciated catchments, to investigate the erosional impact of this glaciation.

Erosion rates range between 30 and 100 mm/ka, comparable to rates measured in non-glaciated regions elsewhere in the Black Forest [1, 2]. Erosion rates of catchments with exposed crystalline basement scale with morphometric measures such as mean catchment slope (Fig. 2). Field studies in tectonically active regions suggest a non-linear relation between mean slope and erosion rate [3]. Our data is best fitted by such a non-linear trend, with a background erosion rate (E_0) of 40 mm/ka and a threshold slope (S_C) of 30°.

Although the landscape of the southern Black Forest was strongly glaciated during the LGM, mean post-glacial erosion rates are comparable to rates measured in non-glaciated regions.

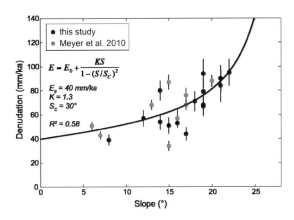

Fig. 2: *Slope vs. catchment-wide erosion rate fitted with a non-linear function proposed by [3], whereas E is the erosion rate, E_0 is the background erosion rate due to chemical weathering, K is a rate constant, S is the mean hillslope angle and S_C is the threshold hillslope angle.*

The correlation between measured erosion rates and morphometric measures suggests that maybe on the catchment-scale erosion is in steady-state with tectonic uplift/base-level lowering along the western border fault of the Black Forest.

[1] H. Meyer et al., Earth Planet. Sci. Lett. 290 (2010) 391

[2] P. Morel et al., Terra Nova 15 (2003) 398

[3] D.R. Montgomery and M.T. Brandon, Earth Planet. Sci. Lett. 201 (2002) 481

[1] *Geology, University of Hannover, Germany*

ANTHROPOGENIC RADIONUCLIDES

http://www.naturagart.de/2012/06/von-ibbenburen-zum-polarkreis/

Anthropogenic iodine-129 in the Arctic Ocean

A first transect of ^{236}U in the Atlantic Ocean

Iodine-129 in German soils

Determination of plutonium in soils

Adsorption of ^{242}Pu(VI) on different clays

^{129}I from a lead target

Long-lived halogen radionuclides in LBE

ANTHROPOGENIC IODINE-129 IN THE ARCTIC OCEAN

Distribution of ^{129}I in the depth profiles and surface water of the Arctic

M. Raiwa[1], A. Daraoui[1], M. Schwinger[1], M. Gorny[1], C. Walther[1], M. Christl, C. Vockenhuber, H.-A. Synal

^{129}I is currently in disequilibrium in all Western European environmental compartments. It has been observed to accumulate in the ecologically sensitive waters of the Arctic Ocean, mainly as a consequence of ^{129}I releases from the European reprocessing plants.

Water samples were collected in 5 depth profiles down to 4238 m in the Arctic Ocean (Eurasian Basin (86° 14.39'N; 59° 17.62'E), North Pole (89° 57.84'N; 150° 49.07'E), Makarov Basin (87° 17.07'N; 165° 15.97'W), Canada Basin (83° 1.80'N; 130° 2.34'W) and Amundsen Basin (83° 19.25'N; 125° 9.99'E)). The sampling campaign (ARK-XXVI/3) was performed by the Alfred Wegener Institute, Bremerhaven, Germany, on the research ship *Polarstern 78* in the expedition *TransArc* during August-September 2011 (Fig. 1).

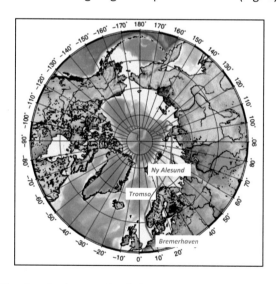

Fig. 1: *Cruise track of the Polarstern expedition ARK-XXVI/3.*

The samples were analyzed for ^{129}I with AMS and for ^{127}I with inductively coupled plasma mass spectrometry (ICP-MS). Details for sample preparation can be found in [1].

The ^{127}I concentrations are fairly constant with depth at all stations (35 – 41 ng/g) while the ^{129}I

concentrations at 800 m depth are about two to three times higher in the Eurasian Basin (6.6 µBq/l) and at the North Pole (5.1 µBq/l) than in the Makarov Basin (2.3 µBq/l) and Amundsen Basin (3.0 µBq/l) (Fig. 2). The lowest value for the surface waters was observed inside the Canada Basin (0.4 µBq/l) corresponding to a ^{129}I/^{127}I ratio of 13x10^{-10}. This considerably exceeds the natural ratio (^{129}I/^{127}I=1.5x10^{-12} [2]). ^{129}I concentrations and consequently ^{129}I/^{127}I ratios at the surface are higher by a factor of >10 than at depth. To explain our results we will have to investigate factors such as the influence of time delay in the transport of ^{129}I from the reprocessing facilities Sellafield and La Hague, dilution by melt ice and river water, and the currents of the Arctic Ocean [3].

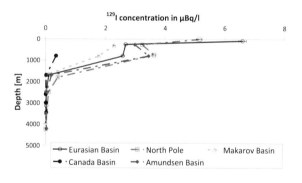

Fig. 2: *^{129}I concentrations in the five depth profiles from the central Arctic.*

[1] R. Michel et al., Sci. Total Environm. 419 (2012) 151

[2] J.E. Moran et al., Chem. Geology 152 (1998) 193

[3] V. Alfimov et al., Mar. Pol. Bull. 49 (2004) 1097

---

[1] *Institute for Radioecology and Radiation Protection, University of Hanover, Germany*

A FIRST TRANSECT OF ^{236}U IN THE ATLANTIC OCEAN

Testing the potential of ^{236}U as a new oceanographic tracer

N. Casacuberta[1], M. Christl, J. Lachner, M. Van-der-Loeff[2], P. Masqué[1], H.-A. Synal

New developments in low energy accelerator mass spectrometry (AMS) allow conducting low-level determinations of heavy ions (including the actinides) at unprecedented sensitivity. In this context, anthropogenic ^{236}U ($T_{1/2}$= 23·10^6 a) has the potential to become a new conservative and transient tracer in oceanography [1-3]. It could be used as a tool to quantify oceanic processes such as water mass mixing and deep-water formation rates on decadal timescales. The presence of ^{236}U in the oceans is mainly due to anthropogenic input, either by nuclear reactor accidents, direct discharges of radioactive waste or the atmospheric nuclear weapons tests performed in the 1950s and 1960s [3].

In this study, ^{236}U/^{238}U ratios and ^{236}U concentrations were determined in 90 seawater samples (3 l each), which were collected from 9 depth profiles during the first two legs of the Dutch GEOTRACES cruise GA02 in 2010 (Fig. 1) along the northwest Atlantic Ocean (from equator to 64ºN).

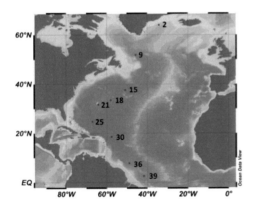

Fig. 1: *Location and station numbers of the samples collected during GA03.*

The measured ^{236}U/^{238}U ratios range from (44±15)·10^{-12} to (1477±91)·10^{-12} (Fig. 2). Higher values correspond to northern water masses (i.e. Labrador Sea Water (LSW) and Northeast

Atlantic Deep Water (NEADW)), which have higher input from global fallout and possibly an additional contribution from the European reprocessing plants La Hague (F) and Sellafield (GB). Lower values are representative of southern water masses such as Antarctic Bottom Water (AABW) and Antarctic Intermediate Water (AAIW).

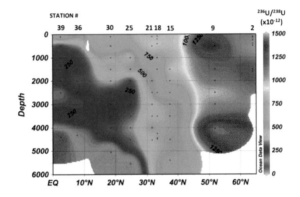

Fig. 2: *Transect of ^{236}U/^{238}U ratio in the North Atlantic Ocean.*

Samples from the Arctic Ocean and the Southern Ocean are currently being processed in order to build a more comprehensive dataset of ^{236}U in oceans. Further work will also involve the measurement of other anthropogenic tracers (e.g. ^{129}I and Pu isotopes) to explore the potential of using the ratio ^{236}U/^{129}I or ^{236}U/^{x}Pu as tracers for different oceanic processes.

[1] Christl et al., Geochim. Cosmochim. Acta 77 (2012) 98

[2] Sakaguchi et al., Earth and Planet. Sci. Lett. 333-334 (2012) 165

[3] Steier et al., Nucl. Instr & Meth. B 266 (2008) 2246

[1] *Environmental Sciences, Universitat Autònoma de Barcelona, Spain*
[2] *Alfred Wegener Institute, Bremerhaven, Germany*

IODINE-129 IN GERMAN SOILS

Values decrease with depth and also vary with the regions

M. Schwinger[1], A. Daraoui[1], B. Riebe[1], C. Walther[1], C. Vockenhuber, H.-A. Synal

Soil samples were taken all over Germany to cover different locations and soil types (Fig. 1).

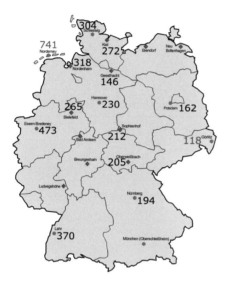

Fig. 1: ^{129}I *deposition densities in mBq/m² for 0 - 50 cm.*

Grassland plots that were left undisturbed for more than 20 - 30 years were selected to sample uninfluenced depth profiles (Fig. 2) with sampling depths of 0-5, 5-10, 10-20, 20-30, and 30-50 cm.

Fig. 2: *Sampling of the grassland plots.*

Iodine was extracted using a dry combustion method [1]. The $^{129}I/^{127}I$ isotopic ratios were determined at the TANDY AMS facility with high efficiency thanks to helium stripping into charge state of 2+.

The measured ^{129}I content decreases rapidly with increasing depth (Fig. 3). Most of the ^{129}I is found in the first 10 cm as reported in the literature [2].

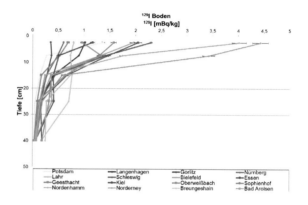

Fig. 3: *Depth profiles of* ^{129}I *in soil samples.*

The figure also indicates that the highest values are found in the northern and western parts of Germany. This can also be seen from the deposition densities (Fig. 1) which are calculated from the ^{129}I content and the corresponding soil density. The deposition densities for the top 50 cm show the highest values for Norderney (north) and Essen (west). The lowest value is found in Görlitz (east). The reason for the high values in the northern and western parts of Germany can most likely be attributed to an input of marine aerosols via rainwater.

[1] A. Daraoui et al., J. Environm. Radioactivity 112 (2012) 8

[2] M. Luo et al., J. Environm. Radioactivity 118 (2013) 30

[1] *Radioecology and Radiation Protection, University of Hannover, Germany*

DETERMINATION OF PLUTONIUM

Pu-isotopes in samples from Fukushima, Chernobyl, and IAEA materials

S. Schneider[1], S. Bister[1], M. Christl, R. Michel[1], V. Schauer[2], G. Steinhauser[2], C. Walther[1]

Plutonium in the environment has primarily been produced by human activities. A large amount was released during the atmospheric nuclear weapons tests. Additional sources of plutonium are reprocessing plants for nuclear fuel and accidents in nuclear facilities. Every source has its own specific isotopic composition. It is therefore possible to identify the origin of the plutonium by measuring the ^{240}Pu/^{239}Pu isotopic ratio. Typical values range from 0.18 for plutonium from atmospheric fallout (from nuclear explosions) to 0.41 typical for fuel from commercial light-water reactors after high burn-up. Due to similar α particle energies in the Pu isotope decays only the sum activity of both isotopes is measured by alpha spectroscopy. AMS is applied for obtaining the isotopic ratio.

In March 2011, the nuclear power plant Fukushima Dai-Ichi was seriously damaged by a tsunami [2]. In order to investigate the potential release of nuclear fuel, soil and plant samples were taken in the vicinity of the damaged plant and investigated with respect to plutonium isotopic ratios.

For validation purposes, soil samples from the areas near Chernobyl and reference samples of the IAEA were also investigated.

The samples were dissolved using nitric and fluoric acid. Subsequently, plutonium was chemically separated from the rest of the nuclides using extraction chromatography [1].

The measured ^{240}Pu/^{239}Pu ratios are shown in Fig. 1. The highest isotopic ratios were found in samples from Ukraine, which were directly contaminated by fallout from the Chernobyl accident. The IAEA samples show results within the expected range (namely 0.26±0.02 for IAEA-375 and 0.19±0.01 for IAEA-Soil-6). The lowest isotopic ratio, which is typical for Pu from detonations, was found in IAEA-sample 384. The results found correspond to the reference values and validate the method.

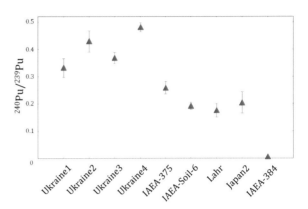

Fig. 1: *Isotopic ratios of the analyzed samples.*

One of the Japanese soil samples exhibits an isotopic ratio of 0.205±0.039, which corresponds to the value for atmospheric fallout. For the other samples it was not possible to determine a ratio, because the measured values lay below the detection limit for ^{240}Pu/^{242}Pu.

Plant samples collected at a distance of 0.88 km and 16 km from the reactor had ^{240}Pu/^{239}Pu isotopic ratios of 0.38±0.05 and 0.64±0.37, respectively. These values agree with the typical ^{240}Pu/^{239}Pu ratio of 0.41 for light-water-reactors like Fukushima after high burn-up.

[1] T. Bisinger et al., Nucl. Instr. & Meth. B 268 (2010) 1269

[2] L. Mohrbach, atw-Intern. J. for Nucl. Power 56 Issue 4/5 (2011) 3

---

[1] *Radioecology and Radiation Protection , University of Hannover, Germany*
[2] *Atomic and Subatomic Physics, Vienna University of Technology, Austria*

ADSORPTION OF ^{242}Pu(VI) ON DIFFERENT CLAYS

Determination of K_D-values at ultra-trace levels

B.-A. Dittmann[1], C. Marquardt[2], M. Christl, H.-A. Synal,

The physical and chemical properties of host rocks are crucial for the safe confinement of radioactive wastes. Based on current knowledge, the following host rocks are the most promising for nuclear waste repositories: rock salt (halite), crystalline rocks (e.g. granite) and clay (respectively shale).

Within the scope of the venture "Retention of repository-relevant radionuclides in natural clay and in saline systems" by the German Federal Ministry of Economics and Technology, a research project has been assigned to the *Institute for Nuclear Waste Disposal* at the *Karlsruhe Institute of Technology* (KIT) to study the speciation of actinides and long-lived fission products in different clay samples (here: kaolinite and opalinus clay). In cooperation with KIT, the subject of this study is to determine the distribution coefficient ($K_D = {}^{242}Pu_{ads.}/{}^{242}Pu_{total}$) of Pu on different clays at ultra-trace levels. The K_D-values are important for a comprehensive safety assessment of a nuclear waste disposal and allow the development of an improved surface-complexation model for Pu on clays.

Clay	s/l [g/mL]	^{242}Pu$_{tot}$ [pg]	^{242}Pu$_{ads}$ [pg]	K_D
Opalinus	20	108.8	107.4	0.98
Opalinus	200	108.4	107.4	0.98
Kaolinite	20	1.35	1.37	1.01
Kaolinite	200	1.35	1.22	0.91
Opalinus	20	1.35	1.16	0.86
Opalinus	200	1.35	1.29	0.96

Tab. 1: *Clay types, s/l = solid(clay)/liquid ratio, total amount of ^{242}Pu added, measured ^{242}Pu adsorbed on clay, and calculated K_D-values.*

For the purpose of the study the clay samples were suspended in synthetic pore water. Then,

≈100 pg or ≈1 pg ^{242}Pu(VI) were added using solutions with different Pu concentrations (Tab. 1). After an equilibration time of 14 days the phases were separated and the solid phase was radiochemically processed to separate the Pu. K_D-values were calculated from the adsorbed mass of ^{242}Pu (measured with the compact 0.6 MV AMS system TANDY) and the total amount of ^{242}Pu added to the sample (Tab. 1).

The results show that under the selected environmental conditions ^{242}Pu(VI) is almost completely absorbed by kaolinite and opalinus clay. However it should be noted that the calculated K_D-values have uncertainties of 13 %. This is mainly because the uncertainty of the initial ^{242}Pu content in the parent solution used for the experiment is systematically influenced by differing results of different ^{242}Pu activity measurements.

The methodology proved to be feasible under the given conditions. For future measurements, however, it will be important to minimize the uncertainty of the total ^{242}Pu mass added to the sample. It is also planned to investigate the amount of plutonium in the solution, as well as the dependency of the oxidation state regarding the adsorption behavior.

[1] B.-A. Dittmann, Diploma thesis (2013) University of Cologne, Germany

[2] M.H. Bradbury et al., Geochim. Cosmochim. Acta 69 (2005) 875

[3] L. K. Fifield, Quat. Geochronol. 3 (2008) 276

[1] *Nuclear Chemistry, University of Cologne, Germany*
[2] *Nuclear Waste Disposal, Karlsruhe Institute of Technology, Karlsruhe, Germany*

^{129}I FROM A LEAD TARGET

Analytics of ^{129}I produced by proton irradiation in a lead target at SINQ

T. Lorenz[1], D. Schumann[2], Y. Dai[2], C. Vockenhuber

At the Paul Scherrer Institut (PSI) spallation neutron source SINQ, a 590 MeV proton beam with a beam current of 2.3 mA is completely stopped in a solid lead target to produce neutrons for the materials sciences, basic physics and particle science. After two years of operation and a total proton dose of more than 10 Ah the target has become highly radioactive. Safety-relevant volatile isotopes and long-lived nuclides are a basic topic for the complete description of the radionuclide inventory. Isotopes in the mass regions below the matrix element are formed by spallation, fragment-ation and high-energy fission processes.

The 0.6 MV TANDY AMS facility was used to determine the content and longitudinal distribution of ^{129}I ($t_{1/2}$ = 1.57x10^7 a) in several pieces along a lead rod of SINQ-Target 4 [1] which was irradiated in 2000/2001.

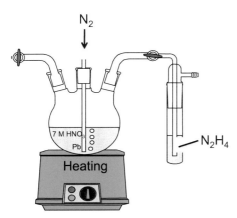

Fig. 1: *Schematic view of the dissolution set-up.*

Each sample material and 10 mg of inactive iodine carrier was dissolved in hot 7 M nitric acid under bubbling nitrogen (Fig. 1). Iodine was separated from the matrix by distillation into hydrazine solution. The iodide was then precipitated with AgNO$_3$ in diluted nitric acid as AgI. After drying at 80°C for several days, the yield was determined gravimetrically (47-77 %).

The material was then further diluted with solid AgI (1:2500) to be suitable for AMS.

Sam.	dist. [mm]	p dose [10^{25}/m^2]	n dose [10^{25}/m^2]	^{129}I [Bq/g]
D03	1.5	5.2	9.2	0.4(1)
D09	25	3.0	7.6	0.5(2)
D10	27	2.7	7.4	0.24(8)
D12	47	1.0	5.7	0.12(2)
D14	49.5	0.8	5.5	0.11(1)

Tab. 1: *Sample parameters: distance from beam center, calculated proton and neutron doses, mean ^{129}I activity concentration per gram lead.*

The 129I activity seems to correlate well with the proton dose (Tab. 1). Apparently, iodine did not evaporate from the lead phase during operation despite the high operation temperatures of up to 425 °C [1]. Previous studies showed that iodine remains dissolved in lead-based alloys up to 500 °C [3]. However, at the moment it is not clear whether iodine is produced in an elementary or ionic state.

Diluting the sample material with solid AgI produced inhomogeneous AMS targets leading to variable ^{129}I activities with rather high uncertainties for the mean data. This could be improved by dilutiing in the liquid phase before precipitating.

[1] Y. Dai et al., J. Nucl. Mat. 343 (2005) 33
[2] T. Lorenz et al., PSI-LCH Annual Report 2012
[3] J. Neuhausen et al., Radiochim. Acta 94 (2006) 239

[1] *Radio- and Environmental Chemistry, PSI, Villigen and Chemistry & Biochemistry, University of Bern*
[2] *Radio- and Environmental Chemistry, PSI, Villigen*

LONG-LIVED HALOGEN RADIONUCLIDES IN LBE

First results for the distribution of [129]I in MEGAPIE samples

B. Hammer[1], A. Türler[1], D. Schumann[2], J. Neuhausen[2], M. Wohlmuther[2], C. Vockenhuber

Lead Bismuth Eutectic (LBE) is an interesting material as it could possibly be used as coolant or target material in future nuclear power plants. During a plant's lifespan a wide range of radionuclides including the long-lived [129]I ($t_{1/2}=1.57 \times 10^7$ a) and [36]Cl ($t_{1/2}=3.01 \times 10^5$ a) is produced [1]. Radiochemical analyses of irradiated LBE are thus important to benchmark theoretical predictions as well as estimate safety hazards of future LBE nuclear facilities during and after operation including options for intermediate or final disposal.

LBE has been used in the experimental liquid metal target MEGAPIE for the neutron spallation source SINQ at the Paul Scherrer Institut (Fig. 1). After this target was removed, samples were taken [2] to chemically separate the halogen radionuclides [129]I and [36]Cl for analysis with AMS. [129]I was measured at the 0.6 MV TANDY AMS facility, while [36]Cl will be measured with the 6 MV EN TANDEM in 2013.

Fig. 1: *Schematic view of the MEGAPIE target.*

Iodine and chlorine were separated from the LBE by distillation. For this purpose, LBE (12 mg), iodine carrier (about 12 mg) and chlorine carrier (10 mg) were dissolved in 5 ml 7 M HNO_3 in a three-neck-flask in a N_2 atmosphere at room temperature. After completely dissolving the LBE, the reaction mixture was heated to 100 °C and iodine and chlorine were distilled into an aqueous hydrazine solution. The hydrazine solution was acidified with 7 M HNO_3 and $AgNO_3$ was added, forming a white precipitate of AgI and AgCl. The precipitate was filtered and redissolved in NH_3 (25%). AgCl is soluble in NH_3,

whereas AgI is insoluble. AgI was filtered and washed 3 times with 5 ml 7 M HNO_3 and dried for about 24 hours at 80 °C. The AgCl fraction was acidified with 7 M HNO_3 and AgCl was re-precipitated and washed with bi-distilled H_2O. The chlorine samples were dried for about 24 hours at 80 °C.

Sample	[129]I [Bq/g]	Sample	[129]I [Bq/g]
H02-U2_St1	2.12E-7	H04-D	1.04E-7
H02-U2_St2	1.85-8	H05-U2-b	1.24E-7
H02-D2	7.52E-8	H05-D3-b	3.66E-8
H03-U6	4.96E-7	H05-B-b	2.20E-6
H03-U12	3.11E-7	H05-D22	1.51E-8

Tab. 1: *[129]I activity concentrations from a small set of samples. Uncertainties of about 10 % result mainly from the final dilution step.*

The measured activity concentrations of [129]I (Tab. 1) are 3 to 6 orders of magnitude lower than those predicted by calculations [3]. At the moment we have no explanation for this discrepancy. Due to its volatility, iodine could have been evaporated from the LBE. However, this seems unlikely because iodine starts to evaporate from LBE at about 500 °C whereas the operating temperature of the MEGAPIE target was less than 340 °C.

[1] A.Y. Konobeyev et al., Nuclear Instr. & Meth. A 605 (2009) 224

[2] B. Hammer et al., PSI Annual Report (2011) 52

[3] L. Zanini et al., PSI Report nr. 08-04 (2008)

[1] *Radio.- and Environmental Chemistry, PSI, Villigen and Chemistry and Biochemistry, University of Bern*
[2] *Radio.- and Environmental Chemistry, PSI, Villigen*

MATERIALS SCIENCES

Gas ionization chamber simulation

MC simulation for the capillary microprobe 1

MC simulation for the capillary microprobe 2

MC simulation for the capillary microprobe 3

Capillary microbeam diameter measurements

Characterization of a position sensitive detector

First PIXE image with a capillary microprobe

Helium channeling in lutetium manganate

He implantation induced swelling in ODS steels

Oxygen stoichiometry of zirconia films

TRIDYN simulation of metal ion etching

Oxygen diffusion profiles in AlSn-nitride films

GAS IONIZATION CHAMBER SIMULATION

Description of an extended Mathematica package

A.M. Müller, M. Döbeli

The Mathematica® package introduced last year for ERDA gas ionization detector simulation [1] was extended to simulate single projectiles with SRIM. A new graphical user interface enables comfortable visualization of selected events of interest (Fig. 1).

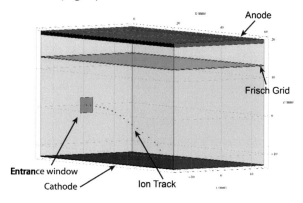

Fig. 1: *Schematic 3D view of the ETHZ gas ionization chamber with the track of a 2 MeV iodine projectile entering the detector filled with 12 mbar isobutane.*

Various detector parameters such as bias voltage and gas pressure can be adjusted dynamically, for which the physical SRIM tables (e.g. range, specific energy loss, etc.) are then rescaled automatically online. The movement of the ionization electrons with time can be illustrated. The resulting signal at the anode can now include the electrostatic shielding inefficiency of the Frisch grid (Fig. 2). Other detector designs, such as Bragg geometries, can also be investigated and illustrated (Fig. 3). This way the influence of different chamber designs on the detector output signal can be examined.

Another intention to extend the simulation package is to gain a better insight into detector pulse form generation. This could improve detector design and help to develop new ideas for an optimized implementation of digital pulse processing algorithms [2].

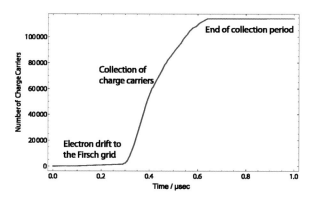

Fig. 2: *Number of collected charge carriers at the anode (corresponding to the preamplifier output signal) for the iodine event illustrated in (Fig. 1). A grid electrostatic shielding inefficiency of 4 % was assumed.*

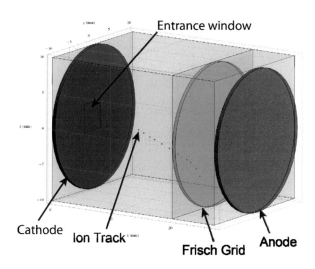

Fig. 3: *Perspective view of the same iodine event as shown in (Fig. 1), but now in a Bragg detector design.*

[1] Laboratory of Ion Beam Physics Annual Report (2011) 35

[2] Laboratory of Ion Beam Physics Annual Report (2012) 16

MC SIMULATION FOR THE CAPILLARY MICROPROBE 1

Description of a 3D Monte Carlo simulation based on SRIM

M.J. Simon, M. Döbeli

Glass capillaries are an easy tool to produce a low current microbeam because the beam spot size is determined by the small outlet diameter of the capillary. The focusing properties of tapered capillaries applied to MeV ions are described in the literature (e.g. [1]). Careful transmission measurements performed at our laboratory with in-house made capillaries showed only decent enhancement factors between 1 and 2, in agreement with recent publications of other research groups (e.g. [2]) but in sharp contrast to [1]. At present, the origin of this small enhancement is still unknown. One hypothesis ascribes this effect to simple elastic scattering of the ions in the glass of the capillary [2]. To test this assumption a 3D Monte Carlo simulation based on SRIM was performed as described below.

Since our capillaries have a regular conical shape, the capillary tip in the simulation is treated as a perfect cone. For the sake of simplicity, surface roughness and penetration of the ions through the capillary wall at the very tip are neglected. Additionally, the ion beam is simulated as perfectly uniform and collimated. Ions scattered from one section of the capillary to the opposite wall are discarded because a second scattering in the direction of the capillary tip is very unlikely.

First, on the order of 10^4 random ion trajectories in borosilicate glass were calculated by SRIM and saved in the so-called EXYZ-file. In the main simulation routine, written in Mathematica®, the entry points of the ions were homogeneously distributed by random numbers over the capillary entry surface. Intersection points of the ions with the cone wall were then calculated and at each of these points one of the stored SRIM trajectories was attached (point A in Fig. 1). If the trajectory intersected the capillary wall again the ion continued on a straight path. Ions which left the capillary through its outlet (Fig. 2) hit a simulated detector at some distance.

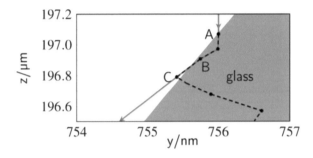

Fig. 1: *Path of an ion (orange) in the capillary. At point A the ion enters the glass wall of the capillary and the SRIM trajectory is attached (dashed line). At B the ion is scattered to C and continues on a straight path.*

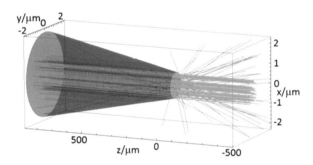

Fig. 2: *Ion trajectories near the capillary tip. Some ions can leave directly the capillary (green), some by scattering (orange).*

At the detector, the impact coordinates and the energy of the particles were recorded. The results of this simulation can be compared to our measurements [3].

[1]   T. Nebiki et al., J. Vac. Sci. & Techn. A21 (2003) 1671

[2]   J. Hasegawa et al., Nucl. Instr. & Meth. B 266 (2008) 2125

[3]   M.J. Simon et al., Laboratory of Ion Beam Physics Annual Report (2012) 76 and 77

MC SIMULATION FOR THE CAPILLARY MICROPROBE 2

Comparison of ion transmission measurements with MC simulation

M.J. Simon, M. Döbeli

Three types of measurements were made to investigate the ion transmission properties of a capillary: (1) rocking curves, i.e. the ion transmissions through capillaries of different taper angles were measured as a function of tilt angle (angle between capillary and beam axis); (2) the energy spectrum of the transmitted ions was measured at each tilt angle; (3) the enhancement factor (ratio of output to input current density) was determined for a perfectly aligned capillary. For these measurements a homogeneous and well collimated ion beam is essential to avoid accidental focusing of the ion beam into the capillary. The results of these measurements can then be compared to 3D MC simulations [1] based on SRIM.

In Fig. 1, a measured rocking curve is compared to simulations of different capillary taper angle Θ. Geometrical parameters such as length or outlet diameter of the capillary were kept unchanged. The measured data is fitted best by the simulation with a taper angle of 0.235°, which agrees within 20% with the optically measured value.

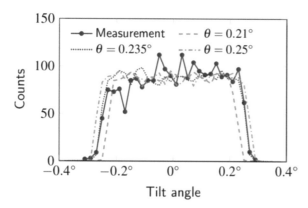

Fig. 1: *Measured and simulated rocking curves.*

In Fig. 2, the measured and simulated energy spectra of a capillary well aligned to the ion beam are compared. They agree quite well but the low energy tail is underestimated in the

simulation, probably due to the simplifications incorporated in the model.

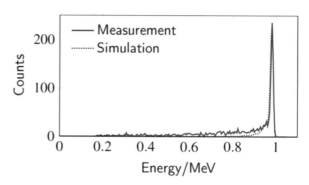

Fig. 2: *Measured and simulated energy spectra of a capillary which was well aligned to the ion beam.*

Fig. 3 shows a compilation of measured and simulated enhancement factors for a number of capillaries.

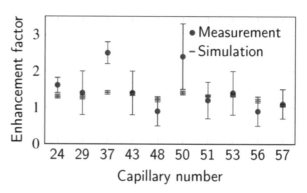

Fig. 3: *Measured and simulated enhancement factors.*

Overall, there is a good agreement between simulation and measurement. This confirms that at least for MeV ions the transmission properties of capillaries can be described well by classical elastic scattering.

[1] M.J. Simon et al., Laboratory of Ion Beam Physics Annual Report (2012) 75

MC SIMULATION FOR THE CAPILLARY MICROPROBE 3

Comparison of phase space measurements with MC simulation

M.J. Simon, M. Döbeli, A.M. Müller, A. Cassimi[1], H. Shiromaru[2], C.L. Zhou[1]

The phase space of 71 MeV Xe^{19+} ions transmitted through glass capillaries of 55 and 50 µm outlet diameter was measured at the IRRSUD beam line of the GANIL facility (Caen, France) [1]. The energy of the transmitted particles and their angle to the primary beam direction was observed with a position sensitive detector.

As explained in [1], the transmitted beam consists of two parts, a core and a halo (Fig. 1). The particles in the core retained the full beam energy whereas the halo particles have suffered energy loss (Fig. 3).

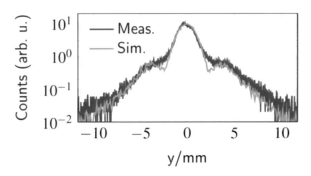

Fig. 2: *Comparison of simulated and measured angular distributions (cut through 2D image).*

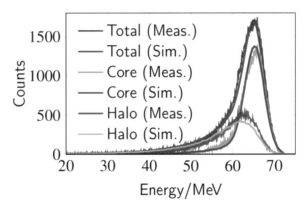

Fig. 3: *Comparison of measured and simulated energy spectra for core and halo particles.*

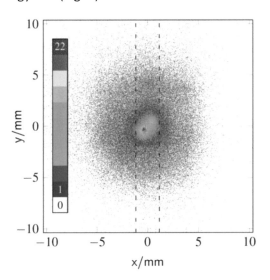

Fig. 1: *Measured 2D image of the 55 micron capillary which was well aligned to the beam. The dashed lines mark the cut used for Fig. 2.*

To test whether elastic scattering can explain these observations, a 3D Monte Carlo simulation [2] was run with the geometrical parameters of the capillaries used. Comparisons between measurement and simulation are shown in Fig. 2 for a cut through the 2D image and in Fig. 3 for the energy spectra of the core and halo particles.

The agreement between simulation and measurement is very good for the two capillaries, both aligned and tilted. This confirms again that elastic scattering perfectly describes the ion transmission properties of glass capillaries at MeV energies.

[1] M.J. Simon et al., Laboratory of Ion Beam Physics Annual Report (2010) 90

[2] M.J. Simon et al., Laboratory of Ion Beam Physics Annual Report (2012) 75

---

[1] *CIMAP, Caen, France*
[2] *Chemistry, Tokyo Metropolitan University, Japan*

CAPILLARY MICROBEAM DIAMETER MEASUREMENTS

Measurements at the in-air microprobe using nuclear track detectors

M. Schulte-Borchers, A. Eggenberger, M.J. Simon, M. Döbeli

Nuclear track detectors have been widely used over the last several decades for the detection of nuclear particles in research and dosimetry. Since the spatial resolution of particle tracks in track detectors is high and mostly independent of particle energy and type, they offer great potential for precise in-air beam diameter measurements at the capillary microprobe of the Laboratory of Ion Beam Physics. These investigations help to improve the basic understanding of ion transmission through the capillary and evaluate the microprobe's applicability for in-air ion beam analysis.

For this purpose pieces of CR-39 track detector foils were cleaned, cut and irradiated in various distances from the capillary outlet (Fig. 1). The spread of the beam diameter was determined over the whole in-air ion range of helium, iodine and gold ions for several energies of a few MeV.

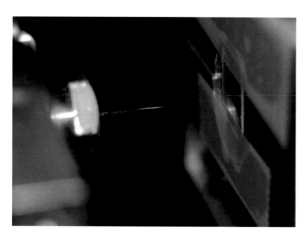

Fig. 1: *Photograph of the irradiation setup of CR-39 with the capillary microprobe.*

After irradiation, following a standard procedure [1] the samples were etched in 6.25 N NaOH at 70 °C for only a few minutes to reveal particle tracks with small diameters. With sufficiently small tracks and a reasonable total dose per detector it is possible to separate most tracks

from each other and thereby determine the ion distribution. From this, a beam diameter - expressed as FWHM of the track density distribution - can be quantified for every single irradiation distance of the capillary outlet.

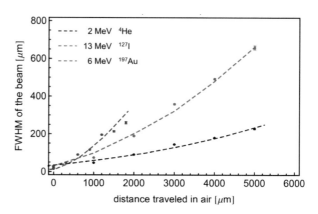

Fig. 2: *Measured ion beam diameters as a function of the distance from the capillary outlet for 2 MeV helium, 13 MeV iodine, and 6 MeV gold ions. The dashed lines show SRIM simulation results for comparison.*

The beam broadening grows significantly as a function of distance travelled in air and increases with the atomic number of the ion as expected due to angular straggling in air. Likewise, measurements of the beam diameter for same ions, but different energies show enhanced straggling for the slower ions.

These trends are in perfect agreement with SRIM simulations (Fig. 2). However, the measured range of ions exiting the capillary was noticeable less. This is due to residual air inside the capillary. The simulations shown in Fig. 2 have therefore been shifted to take this effect into account.

[1] A. Malinowska et al., Nukleonika 53 (2008) S15

CHARACTERIZATION OF A POSITION SENSITIVE DETECTOR

Resolution measurements for a Si PIN diode

M. Schulte-Borchers, A. Eggenberger, M.J. Simon, M. Döbeli

For microcapillaries the dependence of MeV ion transmission on particle mass and energy can be determined with phase space measurements. For this purpose, a position sensitive silicon PIN diode has been characterized to assess its applicability for such measurements.

Fig. 1: *Close-up picture of the Si sensor (edge length: 10 mm) with the microprobe's capillary in close distance.*

Compared to track detectors a position sensitive detector is more useful as additionally it provides the particle energy and it is significantly faster and cheaper to operate in the long term. However, its lateral resolution is typically not as good. Resolution measurements are therefore needed to optimize the design of such phase space experiments.

The sensor (First Sensor DL100-7 CERpin) used here has a 2D readout and thereby provides spatial information additionally to the particle energy (Fig. 1). Data is acquired on four channels with a CAEN digitizer DT5724 in event storage mode. To reduce the time needed for data analysis a new Mathematica® program has been developed to calculate offline both particle energy and location. It is considerably faster

than the online LabVIEW™ based software and can easily analyze the data in between measurements.

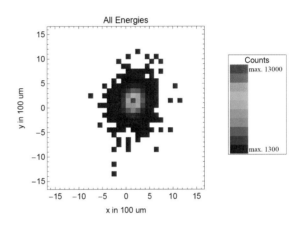

Fig. 2: *Spatial ion distribution of a 5 MeV He irradiation using a round aperture of 250 μm diameter.*

In order to determine energy and position resolution, measurements with circular apertures (diameters between 20 and 500 μm) to precisely define the beam have been made with He particles of 2-5 MeV (Fig. 2). The preliminary FWHM resolution values of the total energy signal are 45 keV (2.3 %) at 2 MeV and 32 keV (0.6 %) at 5 MeV, respectively. The measurements with the smallest apertures yielded preliminary position resolution values of 650 μm at 2 MeV and 250 μm at 5 MeV, respectively.

The position resolution is presently almost as good as expected from the energy resolution and from known data [1]. Optimizing the pulse shaping parameters of the CAEN digital pulse processor for each individual channels might improve the situation.

[1] U. Wahl, SPIRIT Newsletter 4 (2011) http://www.spirit-ion.eu/

FIRST PIXE IMAGE WITH A CAPILLARY MICROPROBE

Two-dimensional elemental distribution on a microscopic scale

M.J. Simon, A. Eggenberger, M. Schulte-Borchers, M. Döbeli

The Laboratory of Ion Beam Physics operates a microprobe based on in-house produced tapered glass capillaries (e.g. [1]). It has been shown in transmission measurements that the capillary mainly acts as a micro-collimator with no significant enhancement of the beam current density [2]. But the extracted current can be increased by focusing the ion beam onto the capillary inlet. Sufficient beam intensity can be obtained this way for applications which require relatively high currents such as PIXE, a method in which the elemental composition of a sample can be determined quantitatively by measuring the characteristic X-ray spectrum.

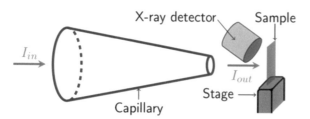

Fig. 1: *Schematic setup of the capillary microprobe for PIXE measurements.*

Fig. 1 shows the schematic setup of the microprobe. The capillary collimates the ion beam to a diameter defined by its outlet size. Because gas leakage through the capillary is low, the ion beam can be extracted into air without a vacuum tight membrane. This simplifies microprobe operation since samples and detectors are easily accessible and electrical charging of samples is avoided. The characteristic X-rays produced in the sample material are detected by a silicon drift diode detector. A two-dimensional image of the elements on the sample surface can be produced by raster scanning the specimen with a piezo-driven XY-stage.

Fig. 2 shows a map of the gold concentration in a gilded microscopy grid. For this measurement

a focused ion beam of 3 MeV protons was collimated by the capillary to a diameter of about 7 µm. The grid was positioned as close as possible to the capillary tip to reduce scattering and energy loss of the ions in air. The grid was rasterscanned in steps of 2.6 µm and at each raster point the number of gold L and M X-rays was measured for 200 ms. For the image about 44'000 points were scanned in 160 minutes.

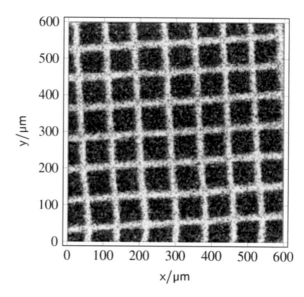

Fig. 2: *Two-dimensional PIXE image of a gold grid (mesh size about 80 µm). The intensity of the gold's M and L X-rays is displayed.*

The grid lines are well resolved and the number of counts per raster point allows for sufficient imaging contrast. The beam current is estimated to be about 4 pA based on the X-ray counting rate. To reduce the measurement time it is planned to increase the beam current by refining the prefocusing technique.

[1]   M.J. Simon et al., Nucl. Instr. & Meth. B 273 (2012) 237

[2]   Laboratory of Ion Beam Physics Annual Report (2012) 75 and 76

HELIUM CHANNELING IN LUTETIUM MANGANATE

C-RBS defect profiles in LuMnO₃ films reveal a sublayer structure

Y. Hu[1], C.W. Schneider[1], M. Döbeli

Crystal defect densities can be measured as a function of depth with Channeling RBS (C-RBS). This technique has been applied to $LuMnO_3$ films epitaxially grown on $YAlO_3$ substrates by Pulsed Laser Deposition [1]. Fig. 1 shows the RBS yield of the Lu signal as a function of the horizontal tilt angle of one of the film towards the beam direction.

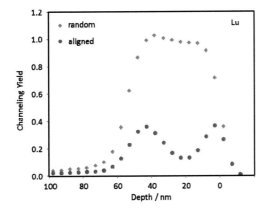

Fig. 1: 2 MeV He C-RBS rocking curve for the Lu signal of a LuMnO₃ film.

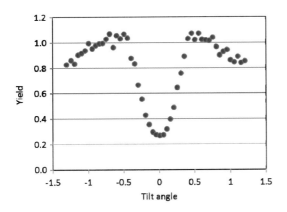

Fig. 2: C-RBS spectra of the Lu signal in aligned and random orientation. The backscattered energy has been converted to a depth scale.

The Channeling RBS energy spectra are shown in random and aligned directions in Fig. 2 for Lu and in Fig. 3 for Mn.

The aligned spectra can be interpreted as defect depth profiles, where the defect density is in arbitrary units. For both constituents, defect-rich regions exist close to the interface and at the surface with an intermediate layer of relatively high crystal quality.

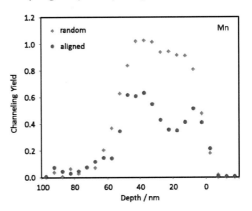

Fig. 3: Channeling RBS spectra for Mn.

The 9 nm thick defect-rich sublayer at the interface can be attributed to the existence of dislocations and local structural incoherence resulting from the accommodation of the lattice mismatch to the substrate. After the partial relaxation of the lattice the film grows for about 40 nm with coherent lattice planes. The approximately 6 nm thick surface layer is partly due to the regular C-RBS surface peak and partly caused by the further lattice relaxation towards the stable bulk structure. This results in the development of enhanced mosaicity at the surface and hence in reduced crystallinity.

The channeling results are in good agreement with XRD and TEM investigations of the same film [1].

[1]   Y. Hu, Ph.D. thesis Nr. 20989 (2013) ETH Zurich

---

[1] General Energy Research, PSI, Villigen

He IMPLANTATION INDUCED SWELLING IN ODS STEELS

Investigation of structural materials for nuclear reactor applications

L. Fave[1], M. Pouchon[1], M. Schulte-Borchers, M. Döbeli

ODS (Oxide Dispersion Strengthened) alloys are of importance for applications in future advanced fission or fusion reactors. Therefore their behavior under intense He irradiation is of special interest.

We have measured the amount of volume swelling of two ODS steels (15CRA-3 and PM2000) induced by He irradiation at temperatures ranging from room temperature to 800 °C in steps of 100 °C. In order to produce a relatively uniform depth profile of radiation damage the particle energy was changed in 7 steps between 800 and 2000 keV for each sample. The fluence was chosen to produce a total of 0.84 dpa (displacements per atom) with a He content of around 4000 atomic ppm/dpa.

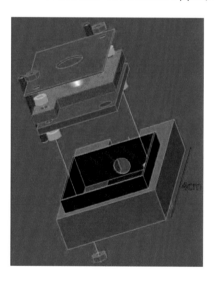

Fig. 1: *Schematics of the furnace used to vary the irradiation temperature.*

The samples were covered by a 400 mesh TEM gilder grid in order to produce a swelling pattern on the surface (Fig. 1). The irradiated and non-irradiated areas could clearly be identified on the samples (Fig. 2). The step profiles were measured by an optical interference microscope

(Fig. 3). From the step heights the volume swelling can be inferred.

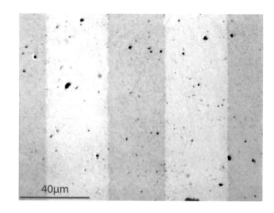

Fig. 2: *Optical microscopy image of a 15CRA-3 sample irradiated at 500 °C.*

Low swelling is found in both alloys. The PM2000 shows a slightly lower resistance against irradiation, with a peak swelling of approximately 0.8 % per dpa at 600 °C, whereas the 15CRA-3 presents an almost constant swelling of around 0.3 % per dpa throughout the temperature range [1].

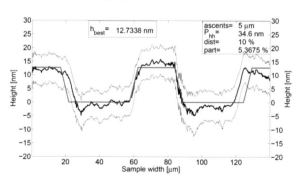

Fig. 3: *Post-processing height profile measured by interference microscopy.*

[1] L. Fave, Master thesis (2012) EPF Lausanne

[1] *Nuclear Energy and Safety, PSI, Villigen*

OXYGEN STOICHIOMETRY OF ZIRCONIA FILMS

The composition of sputter deposited zirconia analyzed with RBS

D. Meier[1], D. Muff[1], R. Spolenak[1], M. Döbeli

The optical properties of reactively sputtered ZrO_x compounds strongly depend on the oxygen stoichiometry of the material. The reduction of the zirconia phases involves the formation of oxygen vacancies and free electrons causing the material to turn from transparent to opaque (Fig. 1). During reactive sputtering, the oxygen content can be controlled by adjusting the oxygen flow rate. While a minimal oxygen flow is needed to produce transparent films, too high reactive gas flows result in low deposition rates due to target poisoning. [1]

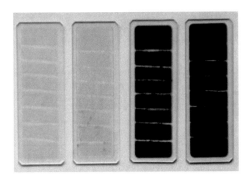

Fig. 1: *Sputter deposited ZrO_x films on glass slides. Oxygen content decreases from left to right.*

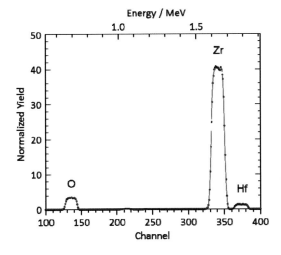

Fig. 2: *RBS spectrum of a 90 nm ZrO_x film grown on a DLC coated silicon substrate.*

In order to find the optimal processing conditions for transparent films with high deposition rates, ZrO_x films were deposited by reactive sputtering under different oxygen flow rates. The composition and thickness of the films were analyzed by 2 MeV Rutherford Backscattering Spectrometry (RBS). In order to facilitate measurements, a batch of samples was deposited on diamond-like carbon (DLC) coated silicon wafers which results in nearly background free RBS spectra that allow very accurate determination of the stoichiometry (Fig. 2). Fig. 3 shows the measured oxygen content as a function of oxygen flow.

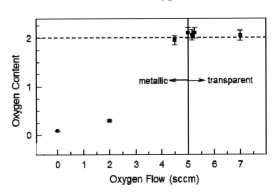

Fig. 3: *Oxygen content x of ZrO_x compounds sputter deposited at different oxygen flow rates.*

Comparison of film thicknesses determined by RBS and profilometry indicated that the films were slightly porous. Ellipsometry was used to measure the refractive index of transparent zirconia films [2]. The findings from this project will be used to produce esthetic coatings on metallic dental implant screws.

[1] S. Berg and T. Nyberg, Thin Solid Films 476 (2005) 215

[2] D. Meier, Master thesis (2013) ETH Zurich

[1] *Nanometallurgy, ETH Zurich*

TRIDYN SIMULATION OF METAL ION ETCHING

Surface layer evolution during cathodic arc zirconium ion etching

J. Ramm[1], B. Widrig[1], A. Dommann[2], X. Mäder[2], A. Neels[2], D. Passerone[3], M. Döbeli

Cathodic arc metal ion etching (MIE) is an efficient technique for in-situ pretreatment of surfaces and influences subsequent layer nucleation. The INNOVA cathodic arc deposition set-up is schematically shown in Fig. 1 with details described in [1].

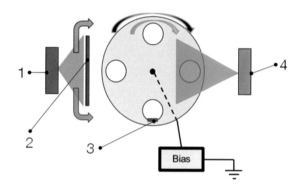

Fig. 1: *Schematics of the INNOVA cathodic arc deposition system. (1) target for MIE, (2) shutter, (3) twofold rotating substrate, (4) target for layer deposition.*

During MIE only ionized metallic vapor produced from the arc source is utilized. The bombardment energy of the ions is controlled by the bias voltage.

Above a certain substrate bias voltage the surface atoms start to be sputtered away, while at the same time the metal ions are implanted deeper into the developing surface layer, eventually leading to an equilibrium intermixed surface composition or to layer growth. This process can be well simulated by the TRIDYN Monte-Carlo program [2].

The evolution of the surface composition during MIE of tungsten carbide (WC) substrates by Zr ions has been investigated by comparing RBS depth profiles to TRIDYN simulations (Fig. 2). Fig. 3 shows the resulting Zr coverage as a function of substrate bias for a constant Zr fluence of approx. $3 \cdot 10^{17}$ cm^{-2}.

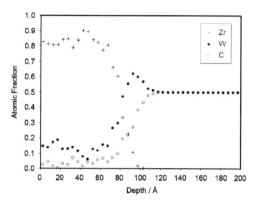

Fig. 2: *Atomic concentration profiles calculated by TRIDYN for 1 keV Zr ions incident on a WC surface. The Zr particle fluence is $8 \cdot 10^{16}$ cm^{-2}.*

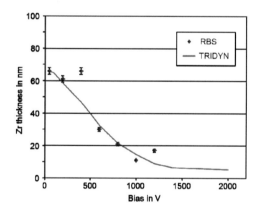

Fig. 3: *Measured and calculated Zr coverage as a function of substrate bias.*

TRIDYN simulations have proven to be a valuable tool in understanding and predicting the MIE process.

[1] J. Ramm, et al., Surf. Coat. Technol. 202 (2007) 876

[2] W. Möller and W. Eckstein, Nucl. Instr. & Meth. B2 (1984) 814

[1] *OC Oerlikon AG, Balzers, Liechtenstein*
[2] *Microsystems Technology, CSEM, Neuchâtel*
[3] *Nanotech@surfaces, Empa, Dübendorf*

OXYGEN DIFFUSION PROFILES IN AlSn-NITRIDE FILMS

Characterization combining light ion RBS and Heavy Ion ERDA

E. Lewin[1], J. Patscheider[1], M. Döbeli

AlN-based nanocomposites are hard and trans–parent coatings where the optical properties can be tuned through the choice and concentration of the alloying element [1, 2]. Thus, this type of material is applicable for multifunctional optical coatings, i.e. scratch resistant optical filters. In the ongoing research at the Laboratory for Nanoscale Materials Science at Empa, different alloying elements are investigated, one of them being tin.

AlSnN films were grown by reactive DC magnetron sputtering in an ultrahigh vacuum chamber. The exact composition of the coating materials and the profile of oxygen diffused into the films were measured by a combination of RBS and Heavy Ion ERDA using 13 MeV [127]I ions. Fig. 1 shows the time-of-flight versus energy raw data, from which the overall content of light elements in the film (Fig. 2) and their concentration depth profiles (Fig. 3) were determined.

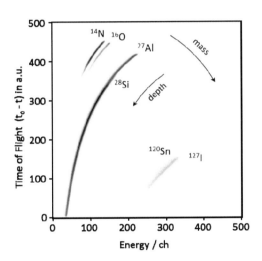

Fig. 1: *Heavy Ion ERDA 2-dimensional spectrum of a 50 nm AlSnN film on silicon.*

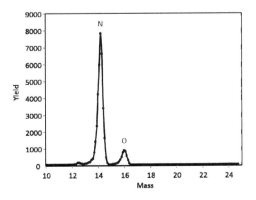

Fig. 2: *Mass spectrum of the film composition extracted from the ERDA data shown in Fig. 1.*

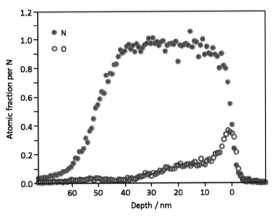

Fig. 3: *Atomic concentration depth profiles of nitrogen and oxygen extracted from the ERDA data shown in Fig. 1.*

The results show that significant oxidation has occurred in films grown at room temperature. Considering the shape of the oxygen profile, it is likely that the oxidation is a result of oxygen diffusion from the surface (which is heavily oxidized), indicating that materials deposited at room temperature are not fully dense.

[1] A. Pélisson et al., Surf. Coat. Technol. 202 (2007) 884

[2] E. Lewin et al., J. Mat. Chem. 22 (2012) 16761

[1] *Nanoscale Materials, Empa, Dübendorf*

EDUCATION

Ion beam analysis laboratory course

One step before university

LIP at LIP

ION BEAM ANALYSIS LABORATORY COURSE

LIP offers IBA tutorial for master students twice a year

M. Döbeli, A.M. Müller, M.J. Simon, M. Schulte-Borchers

Every year in spring and towards the end of the year, the Laboratory of Ion Beam Physics (LIP) organizes a two-day Ion Beam Analysis (IBA) course for master students. The spring course is for ETH students while the fall course is held as part of a master's course of the Department of Chemistry of the University of Bern.

The course starts with a series of introductory lectures on fundamental principles of ion beam physics and the basic IBA techniques RBS, ERDA and PIXE. Then, the students have to prepare samples for the measurement (Fig. 1).

Fig. 1: *One of the tasks is to identify a fake gold coin and determine the actual structure and composition of the material.*

Fig. 2: *Students discuss RBS and PIXE spectra during data taking at the TANDEM laboratory.*

Apart from standard and demonstrative samples, specimens contributed by the students from their own research projects are often included. This considerably adds to the motivation of the participants. RBS and PIXE measurements are then performed at the TANDEM laboratory in groups of 3 to 6 students (Fig. 2). Under the guidance of the LIP staff, the measurements are interpreted afterwards by the participants (Fig. 3).

Fig. 3: *The students have to analyze and interpret their experimental data and answer a number of questions.*

A special analysis procedure has been developed that allows quantifying the experimental data without sophisticated software. Long-term experience has shown that this facilitates the learning process.

Usually about 10 students take part in each course. In many cases course participants have returned in their later career to use IBA techniques at our laboratory.

ONE STEP BEFORE UNIVERSITY

^{14}C dating school project allow a glimpse into interdisciplinary studies

I. Hajdas, C. Biechele, G. Bonani, M. Maurer, H.-A. Synal, L. Wacker

"Studienwoche", the one-week ETH school project was organized in the first week of summer break. During this week, high school students with interests in the natural sciences have a chance to be part of our research group and work on an own research project [1].

Fig. 1: *Sampling wood for ^{14}C dating. Rings were counted for a 'biography' of this tree.*

This year, a team of 5 female students had to reconstruct the time a birch tree from Brig (Canton Valais) had been cut down. This tree has been in store in our laboratory for more than two decades (Fig. 1). The tree rings of this birch recorded environmental conditions that the tree witnessed during its lifetime. The goal of the project was to establish this timeframe.

After counting the rings students took 7 samples starting from the outermost ring. The samples were then prepared for AMS measurements (Fig. 2).

Precise dating is possible for objects made of carbon bearing material that was formed after 1950 AD. The incorporated ^{14}C shows the distinct signal ("bomb peak") of the atmospheric nuclear weapons tests (Fig. 3) [2]. The measured ^{14}C/^{12}C ratio for the outermost ring established that it grew in 1984 AD (Fig. 3). The counted number of rings gave 1858 AD for the first year.

Fig. 2: *Treatment of wood for ^{14}C dating: acid-base-acid (left) prior to graphitization (right).*

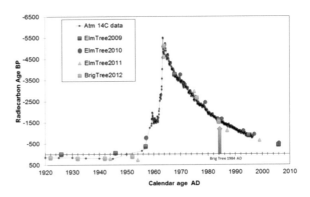

Fig. 3: *Radiocarbon ages of the 7 tree ring samples from the birch tree (blue squares) and results from previous school projects (Elm tree) [1] plotted on the 'bomb peak' curve [2]. The blue arrow shows the age of the outermost ring.*

Like the other school projects before, the 2012 one was successful and an inspiring experience. At least two students confirmed that they would chose physics as their future course of studies.

[1] I. Hajdas et al., Laboratory of Ion Beam Physics Annual Report (2010) 100

[2] I. Levin et al., Sci. Total Environment 391 (2008) 211

LIP AT LIP

School kids reporting on their visit to Ötzi and AMS facility

I. Hajdas, M. Hasler[1]

Last year, school visits at the Laboratory of Ion Beam Physics (LIP) started with a group of children age 8 to 11 from a primary school in Zurich. By coincidence the acronym LIP (Lernen ist Persönlich) of this school is identical to that of our laboratory.

The visit was an interesting and quite inspiring experience for the children, which they documented in their own school photo reportage. Below are some of these reports.

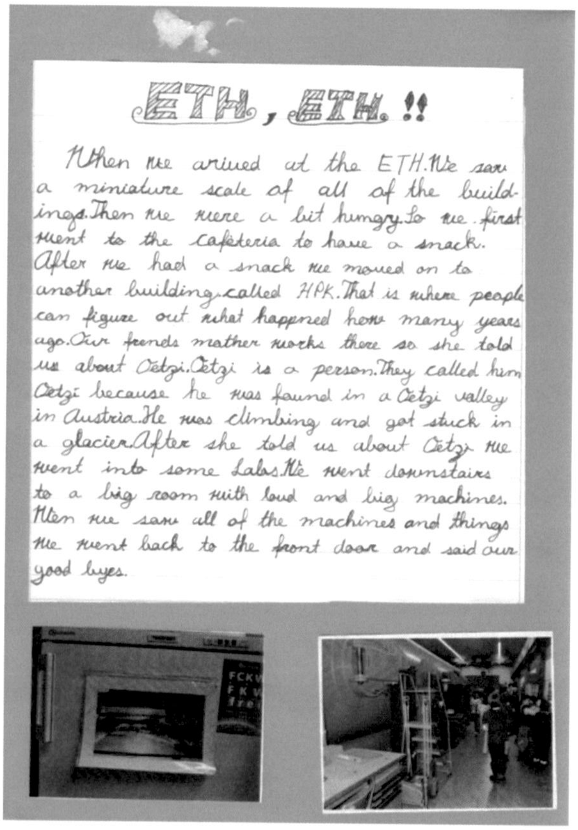

Etienne wrote: "Ötzi is 5000 years old. The ETH has a tiny bit of Ötzi. He was found in Ötzi Tal and frozen. The ETH workers call laboratories "lab". They take small examples of stone and bones. ETH has a good scale".

Others reported on Ötzi as well. Clearly, the very small fragment of the ice mummy kept in our laboratory freezer since the mummy was discovered in 1991 was a special object for the children. Small fragments were radiocarbon dated here to 4550±27 BP [1] which corresponds to a calendar age of 3360-3105 BC. The remaining material is archived (Fig. 1).

Other samples and equipment of our preparatory labs sparked interest and awe. Many details were noticed and documented, some of them showing the funny side of the AMS facility (Fig. 2).

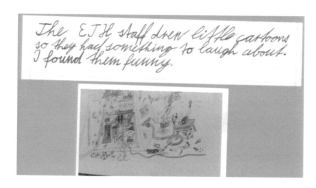

Fig. 2: Fragment of the report written by Léo (10). Picture shows graphic art posted near the EN Tandem accelerator.

The reports written by the LIP kids show their fascination and joy in discovering new things. We hope that their discussions, questions and comments will echo in their future.

Fig. 1: One of the reports written by Charlie (11). Pictures show freezer where fragments of Ötzi tissues are stored (left) and the LIP group visiting the EN Tandem accelerator (right).

[1]   G. Bonani et al., Radiocarbon 36 (1994) 247

[1] *LIP Schule, Zurich; www.lipschule.ch*

PUBLICATIONS

N. Akcar, P. Deline, S. Ivy-Ochs, V. Alfimov, I. Hajdas, P.W. Kubik, M. Christl and C. Schlüchter
The AD 1717 rock avalanche deposits in the upper Ferret Valley (Italy): a dating approach with cosmogenic ^{10}Be
Journal of Quaternary Science **27 (4)** (2012) 383 - 392

N. Akçar, D. Tikhomirov, Ç. Özkaymak, S. Ivy-Ochs, V. Alfimov, H. Sözbilir, B. Uzel and C. Schlüchter
^{36}Cl exposure dating of paleo-earthquakes in the eastern Mediterranean: First results from the western Anatolian Extensional Province, Manisa fault zone, Turkey
Geological Society of America Bulletin **124** (2012) 1724 - 1735

V. Alfimov, A. Aldahan and G. Possnert
Water masses and ^{129}I distribution in the Nordic Seas
Nuclear Instruments and Methods in Physics Research B (2012)
http://dx.doi.org/10.1016/j.nimb.2012.07.042

M. Benz, A. Coskun, I. Hajdas, K. Deckers, S. Riehl, K. W. Alt, B. Weninger and V. Ozkaya
Methodological Implications of New Radiocarbon Dates from the Early Holocene Site of Kortik Tepe, Southeast Anatolia
Radiocarbon **54** (2012) 291 – 304

L. Calcagnile, G. Quarta, L. Maruccio, H.-A. Synal and A.M. Müller
Design features of the new multi isotope AMS beamline at CEDAD
Nuclear Instruments and Methods in Physics Research B (2012)
http://dx.doi.org/10.1016/j.nimb.2012.01.051

M. Christl and P.W. Kubik
New Be-cathode preparation method for the ETH 6 MV Tandem
Nuclear Instruments and Methods in Physics Research B (2012)
http://dx.doi.org/10.1016/j.nimb.2012.03.031

M. Christl, J. Lachner, C. Vockenhuber, I. Goroncy, J. Herrmann and H.-A. Synal
First data of uranium-236 in the North Sea
Nuclear Instruments and Methods in Physics Research B (2012)
http://dx.doi.org/10.1016/j.nimb.2012.07.043

M. Christl, C. Vockenhuber, P.W. Kubik, L. Wacker, J. Lachner, V. Alfimov and H.-A. Synal
The ETH Zurich AMS facilities: Performance parameters and reference materials
Nuclear Instruments and Methods in Physics Research B (2012)
http://dx.doi.org/10.1016/j.nimb.2012.03.004

M. Christl, J. Lachner, C. Vockenhuber, O. Lechtenfeld, I. Stimac, M.R. van der Loeff and H.-A. Synal
A depth profile of uranium-236 in the Atlantic Ocean
Geochimica et Cosmochimica Acta **77** (2012) 98 - 107

X. Dai, M. Christl, S. Kramer-Tremblay and H.-A. Synal
Ultra-trace determination of plutonium in urine samples using a compact accelerator mass spectrometry system operating at 300 kV
Journal of Analytical Atomic Spectrometry **27** (2012) 126 - 130

A. Daraoui, R. Michel, M. Gorny, D. Jakob, R. Sachse, H.-A. Synal and V. Alfimov
Iodine-129, Iodine-127 and Caesium-137 in the environment: soils from Germany and Chile
Journal of Environmental Radioactivity **112** (2012) 8 – 22

L. Di Nicola, C. Baroni, S. Strasky, M.C. Salvatore, C. Schlüchter, N. Akçar, P.W. Kubik and R. Wieler
Multiple cosmogenic nuclides document the stability of the East Antarctic Ice Sheet in northern Victoria Land since the Late Miocene (5-7 Ma)
Quaternary Science Reviews **57** (2012) 85 – 94

F. Donadini, A. Motschi, C. Rosch and I. Hajdas
Combining an archaeomagnetic and radiocarbon study: dating of medieval fireplaces at the Muhlegasse, Zurich
Journal of Archaeological Science **39** (2012) 2153 – 2166

S.M. Fahrni, L. Wacker, H.-A. Synal and S. Szidat
Improving a gas ion source for ^{14}C AMS
Nuclear Instruments and Methods in Physics Research B (2012)
http://dx.doi.org/10.1016/j.nimb.2012.03.037

D. Flak, A. Braun, B.S. Munc, M. Döbeli, T. Graule and M. Rekas
Electronic structure and surface properties of nonstoichiometric $Fe_2O_{3-\delta}$ (α and γ) and its application in gas sensing
Procedia Engineering **47** (2012) 257 - 260

D.R. Griffith, L. Wacker, P.M. Gschwend and T.I. Eglinton
Carbon isotopic (^{13}C and ^{14}C) composition of synthetic estrogens and progestogens
Rapid Communications in Mass Spectrometry **26** (2012) 2619 - 2626

D. Güttler, L. Wacker, B. Kromer, M. Friedrich and H.-A. Synal
Evidence of 11-year solar cycles in tree rings from 1010 to 1110 AD – Progress on high precision AMS measurements
Nuclear Instruments and Methods in Physics Research B (2012)
http://dx.doi.org/10.1016/j.nimb.2012.08.046

N. Haghipour, J.-P. Burg, F. Kober, G. Zeilinger, S. Ivy-Ochs, P.W. Kubik and M. Faridi
Rate of crustal shortening and non-Coulomb behaviour of an active accretionary wedge:The folded fluvial terraces in Makran (SE, Iran)
Earth and Planetary Science Letters **355 - 356** (2012) 187 - 198

I. Hajdas, J. Trumm, G. Bonani, C. Biechele, M. Maurer and L. Wacker
Roman Ruins as an Experiment for Radiocarbon Dating of Mortar
Radiocarbon **54** (2012) 897 – 903

J.R. Hein, T.A. Conrad, M. Frank, M. Christl and W.W. Sager
Copper-nickel-rich, amalgamated ferromanganese crust-nodule deposits from Shatsky Rise, NW Pacific
Geochemistry, Geophysics, Geosystems **13** (2012) 1 - 23

K. Hippe, F. Kober, L. Wacker, S.M. Fahrni, S. Ivy-Ochs, N. Akçar, C. Schlüchter and R. Wieler
An update on in situ cosmogenic ^{14}C analysis at ETH Zürich
Nuclear Instruments and Methods in Physics Research B (2012)
http://dx.doi.org/10.1016/j.nimb.2012.06.020

F. Kober, K. Hippe, B. Salcher, S. Ivy-Ochs, P.W. Kubik, L. Wacker and N. Hählen
Debris-flow–dependent variation of cosmogenically derived catchment-wide denudation rates
Geological Society of America Bulletin **40** (2012) 935 - 938

B. Kromer, S. Lindauer, H.-A. Synal and L. Wacker
MAMS – A new AMS facility at the Curt-Engelhorn-Centre for Archaeometry, Mannheim, Germany
Nuclear Instruments and Methods in Physics Research B
http://dx.doi.org/10.1016/j.nimb.2012.01.015

J. Lachner, M. Christl, H.-A. Synal, M. Frank and M. Jakobsson
Carrier free ^{10}Be/^{9}Be measurements with low-energy AMS: Determination of sedimentation rates in the Arctic Ocean
Nuclear Instruments and Methods in Physics Research B (2012)
http://dx.doi.org/10.1016/j.nimb.2012.07.016

J. Lachner, M. Christl, C. Vockenhuber and H.-A. Synal
Detection of UH^{3+} and ThH^{3+} molecules and ^{236}U background studies with low-energy AMS
Nuclear Instruments and Methods in Physics Research B (2012)
http://dx.doi.org/10.1016/j.nimb.2012.02.010

J. Lachner, M. Christl, C. Vockenhuber and H.-A. Synal
Existence of triply charged actinide-hydride molecules
Physical Review **A 85** (2012) 022717-1 - 022717-6

J. Lachner, I. Dillmann, T. Faestermann, G. Korschinek, M. Poutivtsev, G. Rugel, C.L. von Gostomski, A. Türler and U. Gerstmann
Attempt to detect primordial ^{244}Pu on Earth
Physical Review C **85** (2012) DOI: 10.1103/PhysRevC.85.015801

J. Lippold, S. Mulitza, G. Mollenhauer, S. Weyer, D. Heslop and M. Christl
Boundary scavenging at the East Atlantic margin does not negate use of ^{231}Pa/^{230}Th to trace Atlantic overturning
Earth and Planetary Science Letters **333 - 334** (2012) 317 - 331

B.C. Lougheed, I. Snowball, M. Moros, K. Kabel, R. Muscheler, J.J. Virtasalo and L. Wacker
Using an independent geochronology based on palaeomagnetic secular variation (PSV) and atmospheric Pb deposition to date Baltic Sea sediments and infer ^{14}C reservoir age
Quaternary Science Reviews **42** (2012) 43 - 58

M. Mann, J. Beer, F. Steinhilber and M. Christl
Be-10 in lacustrine sediments-A record of solar activity?
Journal of Atmospheric and Solar-Terrestrial Physics **80** (2012) 92-99

L. C. Malatesta, S. Castelltort, S. Mantellini, V. Picotti, I. Hajdas, G. Simpson, A. E. Berdimuradov, M. Tosi and S. D. Willett
Dating the Irrigation System of the Samarkand Oasis: A Geoarchaeological Study
Radiocarbon **54** (2012) 91 – 105

M. Martschini, P. Andersson, O. Forstner, R. Golser, Dag Hanstorp, A.O. Lindahl, W. Kutschera, S. Pavetich, A. Priller, J. Rohlén, P. Steier, M. Suter and A. Wallner
AMS of ^{36}Cl with the VERA 3 MV tandem accelerator
Nuclear Instruments and Methods in Physics Research B (2012)
http://dx.doi.org/10.1016/j.nimb.2012.01.055

S. Merchel, W. Bremser, S. Akhmadaliev, M. Arnold, G. Aumaître, D.L. Bourlès, R. Braucher, M. Caffee, M. Christl, L.K. Fifield, R.C. Finkel, S.P.H.T. Freeman, A. Ruiz-Gómez, P.W. Kubik, M. Martschini, D.H. Rood, S.G. Tims, A. Wallner, K.M. Wilcken and S. Xu
Quality assurance in accelerator mass spectrometry: Results from an international round-robin exercise for ^{10}Be
Nuclear Instruments and Methods in Physics Research B **289** (2012) 68 - 73

R. Michel, A. Daraoui, M. Gorny, D. Jakob, R. Sachse, L. Tosch, H. Nies, I. Goroncy, J. Herrmann, H.-A. Synal, M. Stocker and V. Alfimov
Iodine-129 and iodine-127 in European seawaters and in precipitation from Northern Germany
Science of the Total Environment **419** (2012) 151 - 169

M. Molnár, I. Hajdas, R. Janovics, L. Rinyu, H.-A. Synal, M. Veres and L. Wacker
C-14 analysis of groundwater down to the millilitre level
Nuclear Instruments and Methods in Physics Research B (2012)
http://dx.doi.org/10.1016/j.nimb.2012.03.038

A.M. Müller, M. Döbeli, M. Suter and H.-A. Synal
Performance of the ETH gas ionization chamber at low energy
Nuclear Instruments and Methods in Physics Research B **287** (2012) 94 - 102

C. Münsterer, J. Fohlmeister, M. Christl, A. Schröder-Ritzrau, V. Alfimov, S. Ivy-Ochs, A. Wackerbarth and A. Mangini
Cosmogenic ^{36}Cl in karst waters from Bunker Cave, northwestern Germany – A tool to derive local evapotranspiration?
Geochimica et Cosmochimica Acta **86** (2012) 138 - 149

E. Nottoli, M. Arnold, G. Aumaître, D.L. Bourlès, K. Keddadouche and M. Suter
The physics behind the isobar separation of ^{36}Cl and ^{10}Be at the French AMS national facility ASTER
Nuclear Instruments and Methods in Physics Research B (2012)
http://dx.doi.org/10.1016/j.nimb.2012.01.052

R. Raghavan, B. Kombaiah, M. Döbeli, R. Erni, U. Ramamurty and J. Michler
Nanoindentation response of an ion irradiated Zr-based bulk metallic glass
Materials Science and Engineering a-Structural Materials Properties Microstructure and Processing **532** (2012) 407-413

J. Rethemeyer, R.-H. Fülöp, S. Höfle, L. Wacker, S. Heinze, I. Hajdas, U. Patt, S. König, B. Stapper and A. Dewald
Status report on sample preparation facilities for ^{14}C analysis at the new Cologne AMS center
Nuclear Instruments and Methods in Physics Research B (2012)
http://dx.doi.org/10.1016/j.nimb.2012.02.012

L. Rinyu, M. Molnár, I. Major, T. Nagy, M. Veres, Á. Kimák, L. Wacker and H.-A. Synal
Optimization of sealed tube graphitization method for environmental C-14 studies using MICADAS
Nuclear Instruments and Methods in Physics Research B (2012)
http://dx.doi.org/10.1016/j.nimb.2012.08.042

I. Röhringer, R. Zech, U. Abramowski, P. Sosin, A. Aldahan, P.W. Kubik, L. Zöller and W. Zech
The late Pleistocene glaciation in the Bogchigir Valleys (Pamir, Tajikistan) based on ^{10}Be surface exposure dating
Quaternary Research **78** (2012) 590 - 597

T. Ryll, P. Reibisch, L. Schlagenhauf, A. Bieberle-Huetter, M. Döbeli, J.L.M. Rupp and L.J. Gauckler
Lanthanum nickelate thin films deposited by spray pyrolysis: Crystallization, microstructure and electrochemical properties
Journal of the European Ceramic Society **32** (2012) 1701 - 1709

I. Schimmelpfennig, J.M. Schaefer, N. Akçar, S. Ivy-Ochs, R.C. Finkel and C. Schlüchter
Holocene glacier culminations in the Western Alps and their hemispheric relevance
Geological Society of America Bulletin **40** (2012) 891 - 894

I. Schindelwig, N. Akcar, P.W. Kubik and C. Schlüchter
Lateglacial and early Holocene dynamics of adjacent valley glaciers in the Western Swiss Alps
Journal of Quaternary Science **27** (2012) 114-124

M.J. Simon, M. Döbeli, A.M. Müller and H.-A. Synal
In-air STIM with a capillary microprobe
Nuclear Instruments and Methods in Physics Research B **273** (2012) 237 - 240

F. Steinhilber, J.A. Abreu, J. Beer, I. Brunner, M. Christl, H. Fischer, U. Heikkila, P.W. Kubik, M. Mann, K.G. McCracken, H. Miller, H. Miyahara, H. Oerter and F. Wilhelms
9,400 years of cosmic radiation and solar activity from ice cores and tree rings
Proceedings of the National Academy of Sciences of the United States of America **109** (2012) 5967-5971

H.-A. Synal, T. Schulze-König, M. Seiler, M. Suter and L. Wacker
Mass spectrometric detection of radiocarbon for dating applications
Nuclear Instruments and Methods in Physics Research B (2012)
http://dx.doi.org/10.1016/j.nimb.2012.01.026

F. Van den Berg, F. Schlunegger, N. Akçar and P. Kubik
^{10}Be-derived assessment of accelerated erosion in a glacially conditioned inner gorge, Entlebuch, Central Alps of Switzerland
Earth Surface Processes and Landforms (2012) DOI: 10.1002/esp.3237

C. Vockenhuber, V. Alfimov, M. Christl, J. Lachner, T. Schulze-König, M. Suter and H.-A. Synal
The potential of He stripping in heavy ion AMS
Nuclear Instruments and Methods in Physics Research B (2012)
http://dx.doi.org/10.1016/j.nimb.2012.01.014

L. Wacker, S.M. Fahrni, I. Hajdas, M. Molnar, H.-A. Synal, S. Szidat and Y.L. Zhang
A versatile gas interface for routine radiocarbon analysis with a gas ion source
Nuclear Instruments and Methods in Physics Research B (2012)
http://dx.doi.org/10.1016/j.nimb.2012.02.009

L. Wacker, R.-H. Fülöp, I. Hajdas, M. Molnár and J. Rethemeyer
A novel approach to process carbonate samples for radiocarbon measurements with helium carrier gas
Nuclear Instruments and Methods in Physics Research B (2012)
http://dx.doi.org/10.1016/j.nimb.2012.08.030

L. Wacker, J. Lippold, M. Molnár and H. Schulz
Towards radiocarbon dating of single foraminifera with a gas ion source
Nuclear Instruments and Methods in Physics Research B (2012)
http://dx.doi.org/10.1016/j.nimb.2012.08.038

L. Wacker, C. Münsterer, B. Hattendorf, M. Christl, D. Günther and H.-A. Synal
Direct coupling of a laser ablation cell to an AMS
Nuclear Instruments and Methods in Physics Research B (2012)
http://dx.doi.org/10.1016/j.nimb.2012.02.013

J. Wambach, M. Schuberta, M. Döbeli and F. Vogel
Characterization of a spent Ru/C catalyst after gasification of biomass in supercritical water
Chimia **66** (2012) 706 - 711

I.G.M. Wientjes, R.S.W. van de Wal, M. Schwikowski, A. Zapf, S. Fahrni and L. Wacker
Carbonaceous particles reveal that Late Holocene dust causes the dark region in the western ablation zone of the Greenland ice sheet
Journal of Glaciology **58** (2012) 787 - 794

C. Wirsig, R.O. Kowsmann, D.J. Miller, J.M. de Oliveira Godoy and A. Mangini
U/Th-dating and post-depositional alteration of a cold seep carbonate chimney from the Campos Basin offshore Brazil
Marine Geology **329 - 331** (2012) 24 - 33

H. Wittmann, F. von Blanckenburg, J. Bouchez, N. Dannhaus, R. Naumann, M. Christl and J. Gaillardet
The dependence of meteoric ^{10}Be concentrations on particle size in Amazon River bed sediment and the extraction of reactive $^{10}Be/^{9}Be$ ratios
Chemical Geology **318 - 319** (2012) 126 - 138

B.M. Wojek, E. Morenzoni, D.G. Eshchenko, A. Suter, T. Prokscha, H. Keller, E. Koller, Ø. Fischer, V.K. Malik, C. Bernhard and M. Döbeli
Magnetism, superconductivity, and coupling in cuprate heterostructures probed by low-energy muon-spin rotation
Physical Review B **85** (2012) DOI: 10.1103/PhysRevB.85.024505

Y.L. Zhang, N. Perron, V.G. Ciobanu, P. Zotter, M.C. Minguillon, L. Wacker, A.S.H. Prevot, U. Baltensperger and S. Szidat
On the isolation of OC and EC and the optimal strategy of radiocarbon-based source apportionment of carbonaceous aerosols
Atmospheric Chemistry and Physics **12** (2012) 10841-10856

T. Zimmerling, K. Mattenberger, M. Döbeli, M.J. Simon and B. Batlogg
Deep trap states in rubrene single crystals induced by ion radiation
Physical Review B **85** (2012) DOI: 10.1103/PhysRevB.85.134101

TALKS AND POSTERS

N. Akçar, C. Özkaymak, D. Tikhomirov, S. Ivy-Ochs, Ö. Sümer, H. Sözbilir, V. Alfimov, B. Uzel and C. Schlüchter
Aktif Normal Faylarda Kozmojenik ^{36}Cl Uygulamalar : Bat Anadolu'dan Ik Sonuçlar
Turkey, Istanbul, 18.-19.10.2012, ATAG 16

N. Bellin, V. Vanacker and P.W. Kubik
Contrasting modern and ^{10}Be-derived erosion rates for the southern Betic Cordillera, Spain
Austria, Vienna, 22.-27.4.2012, EGU General Assembly

R. Belmaker, N. Tepelyakov, J. Beer, P.W. Kubik, M. Christl, B. Lazar and M. Stein
Cosmogenic beryllium isotopes in halite and the age of the Sedom Formation, Israel
France, Paris, 09.07.2012, 21st Radiocarbon Conference

A. Birkholz, M. Gierga, I. Hajdas, R. Smittenberg, L. Wacker and S.M. Bernasconi
Compound-specific radiocarbon dating - a tool for dating lake sediments?
Austria, Vienna, 22.-27.4.2012, EGU General Assembly

J.H. Blöthe, H. Munack, A. Fülling, P.W. Kubik and O. Korup
Late Pleistocene to recent incision dynamics near the western Tibetan Plateau margin, Zanskar, India
Austria, Vienna, 22.-27.4.2012, EGU General Assembly

M. Buechi, F. Kober, S. Ivy-Ochs and P.W. Kubik
Influence of former glaciations on interglacial landscape evolution: A case study from the LGM nunatak Hörnli
Austria, Vienna, 22.-27.4.2012, EGU General Assembly

M. Christl, J. Lachner, C. Vockenhuber and H.-A. Synal
Aktinidenmessungen am ETH Kleinbeschleuniger „TANDY"
Germany, Stuttgart, 16.03.2012, DPG Spring Meeting

A. Claude, S. Ivy-Ochs, F. Kober, M. Antognini, B. Salcher and P.W. Kubik
Geomorphology and landscape evolution at the Chironico landslide (Leventina valley, Swiss Alps)
Switzerland, Berne, 10.03.2012, CH-QUAT Annual Meeting 2012

A. Claude, S. Ivy-Ochs, F. Kober, M. Antognini, B. Salcher, M. Christl and P.W. Kubik
Geomorphology and landscape evolution at the Chironico landslide (Leventina valley, Swiss Alps)
Austria, Vienna, 22.-27.4.2012, EGU General Assembly

A. Claude, S. Ivy-Ochs, F. Kober, M. Antognini, B. Salcher and P.W. Kubik
Geomorphology and landscape evolution at the Chironico landslide (Leventina valley, Swiss Alps)
Switzerland, Berne, 16.-17.11.2012, 10th Swiss Geoscience Meeting

N. Dannhaus, F. von Blanckenburg, H. Wittmann, P. Kram and M. Christl
The ^{10}Be$_{(meteoric)}$/^{9}Be isotope ratio as a new tracer for Earth surface erosion and weathering
Germany, Münster, 01.06.2012, Deutsche Mineralogische Gesellschaft Sektion für Geochemie

N. Dannhaus, F. von Blanckenburg, H. Wittmann, P. Kram and M. Christl
Measuring denudation rates with the $^{10}Be_{(meteoric)}/^{9}Be$ isotope ratio from creek to continental drainage basin scale on contrasting bedrock types
Germany, Hamburg, 23.-28.08.2112, Geologische Vereinigung & Sediment Meeting

M. Döbeli
Ion beam analysis of materials
Switzerland, Villigen, 24.05.2012, Theorie Seminar, Paul Scherrer Institut

M. Döbeli, M.J. Simon, M. Schulte-Borchers and A.M. Müller
A capillary microbeam for STIM and PIXE
USA, Fort Worth, 09.08.2012, Conference on the Application of Accelerators in Research and Industry

E.F. Gjermundsen, J.P. Briner, O. Salvigsen, N. Akcar, P.W. Kubik, N. Gantert and A. Hormes
Late Weichselian ice sheet configuration in northwest Spitsbergen, from ^{10}Be dating and lithological studies of erratic boulders and bedrock.
USA, Winter Park, 07.-09.3.2012, 42nd International Arctic Workshop

L. Grämiger, J.R. Moore and S. Ivy-Ochs
Prehistoric rock avalanches at Rinderhorn, Switzerland
Austria, Vienna, 22.-27.4.2012, EGU General Assembly

D. Güttler, M. Friedrich, B. Kromer, Corina Solis, H.-A. Synal and L. Wacker
High precision ^{14}C AMS-analysis of tree rings - An evaluation of sample preparation and reproducibility of AMS measurements
France, Paris, 09.07.2012, 21st Radiocarbon Conference

N. Haghipour, J-P. Burg, F. Kober, G. Zeilinger, S. Ivy-Ochs, P.W. Kubik and M. Faridi
Rate of crustal shortening and non-Coulomb behavior of an active accretionary wedge: The folded fluvial terraces in Makran (SE, Iran)
Australia, Brisbane, 05.-10.8.2012, International Geological Congress

I. Hajdas, A. Birkholz, C. Biechele, G. Bonani, M. Gierga, M. Maurer, A. Michczynski and L. Wacker
Varve sediments as a tool for development of ^{14}C dating method on ultra-small samples
Germany, Manderschein, 23.-24. 03.2012, Third Workshop of the PAGES Varves Working Group

I. Hajdas, S. F. M. Breitenbach, M. Gierga, G. H. Haug, J. F. Adkins, C. Biechele, G. Bonani, M. Maurer and L. Wacker
^{14}C in a stalagmite from NE India: preliminary results of dating near the limit of radiocarbon time scale
Austria, Vienna, 22.-27.4.2012, EGU General Assembly

I. Hajdas, C. Biechele, G. Bonani, M. Maurer and L. Wacker
Curious and 'not publishable' ^{14}C ages
France, Paris, 08.-13.07.2012, 21st Radiocarbon Conference

I. Hajdas, C. Biechele, G. Bonani, M. Maurer and L. Wacker
Sample specific preparation—overview of treatment methods applied at the ^{14}C laboratory ETH Zurich
France, Paris, 08.-13.07.2012, 21st Radiocarbon Conference

I. Hajdas, S. F. M. Breitenbach, M. Gierga, G. H. Haug, J. F. Adkins, C. Biechele, G. Bonani, M. Maurer and L. Wacker
Testing the potential of a stalagmite from NE India for extension of the ^{14}C calibration curve
France, Paris, 08.-13.07.2012, 21st Radiocarbon Conference

I. Hajdas
From radiocarbon ages to calendar time scale
Poland, Gliwice, 06.-13.9.2012, INTIMATE WG1 Summer Training School on Dating Methods

I. Hajdas
^{14}C dating for reliable chronologies of various archives: from sample treatment to calendar time scale
Greece, Thira, Santorini, 07.-10.10.2012, Earthtime EU

I. Hajdas
Radiocarbon dating of lake sediments for understanding of the past 50,000 years
Hungary, Debrecen, 29.11.2012, ATOMIKI

I. Hajdas
^{14}C calibration curve update
France, Bordeaux, 10.-13.12.2012, ACER-INTIMATE workshop: Harmonisation of Database Chronologies

K. Hippe, F. Kober, S. Ivy-Ochs, L. Wacker, P.W. Kubik and R. Wieler
Quantifying sediment storage using combined cosmogenic in situ ^{14}C-^{10}Be-^{26}Al analysis
Austria, Vienna, 22.-27.4.2012, EGU General Assembly

K. Hippe, F. Kober, S. Ivy-Ochs, L. Wacker, P.W. Kubik, C. Schlüchter and R. Wieler
Combining in situ cosmogenic ^{14}C and ^{10}Be analysis to study complex surface processes and exposure histories
France, Paris, 09.07.2012, 21nd Radiocarbon Conference

K. Hippe, S. Ivy-Ochs, F. Kober, J. Zasadni, R. Wieler, L. Wacker, P. Kubik and C. Schlüchter
Chronology of deglaciation and Lateglacial ice flow reorganization in the Gotthard Pass area, Central Swiss Alps, based on cosmogenic ^{10}Be and in situ ^{14}C
Switzerland, Berne, 06.-17.11.2012, 10th Swiss Geoscience Meeting

A. Hormes, E.F. Gjermundsen, N. Akcar and P.W. Kubik
Achievements and gaps of cosmogenic nuclide dating in the High Arctic
Norway, Trondheim, 21.-22.5.2012; Nordic Workshop on Cosmogenic Nuclide Dating

S. Ivy-Ochs
Dating Trentino landslides with cosmogenic nuclides
Italy, Trento, 06.02.2012, Research Collaboration Meeting

S. Ivy-Ochs, H. Kerschner and C. Schlüchter
Timing of deglaciation in the northern Alps based on cosmogenic nuclide data
Austria, Vienna, 22.-27.4.2012, EGU General Assembly

S. Ivy-Ochs, S. Martin, P. Campedel, V. Alfimov, E. Andreotti, A. Viganò, G. Carugati, C. Vockenhuber and S. Cocco
Constructing a time series for large landslides in Trentino (Italy) with ^{36}Cl exposure dating
Austria, Vienna, 22.-27.4.2012, EGU General Assembly

S. Ivy-Ochs
Surface exposure dating large landslides in the Alps
Switzerland, Zurich, 16.05.2012, AMS Seminar

S. Ivy-Ochs, I. Hajdas, M. Maisch, H. Kerschner, N. Akçar, C. Vockenhuber, P.W. Kubik, C. Schlüchter
A comparison of exposure-dated and radiocarbon-dated Sites in the Alps: What can we learn?
France, Paris, 08.-13.07.2012, 21st Radiocarbon Conference

S. Ivy-Ochs
Constructing cosmogenic nuclide chronologies
Greece, Thira, Santorini, 07.-10.10.2012, Earthtime EU

S. Ivy-Ochs
Cosmogenic nuclides and the timing of Lateglacial glacier variations
Austria, Bludenz, 08.-10.11.2012, Alpine terrestrial records COST-INTIMATE workhop

S. Ivy-Ochs
Climate events at the end of the Last Glacial Maximum
Austria, Vienna, 27.11.2012, Nowagea: Women in Earth Sciences

S. Ivy-Ochs
Using cosmogenic nuclides to date large landslides in the Alps
Austria, Vienna, 28.11.2012, Geosciences Colloquium

S. Ivy-Ochs
Summary of cosmogenic nuclide data for Alpine LGM sites
Italy, Udine, 05.-07.12.2012, Circumalpine Events and Correlations during the Late Pleistocene

J. Jensen, J. Julin, H. Kettunen, M. Laitinen, O. Osmani, M. Rossi, T. Sajavaara, A. Schinner, P. Sigmund, C. Vockenhuber and H.J. Whitlow
Straggling of MeV Kr ions in gases
Japan, Kyoto, 21.-25.10.2012, 25th International Conference on Atomic Collisions in Solids

F. Kober, K. Hippe, B. Salcher, S. Ivy-Ochs, P.W. Kubik and L. Wacker
Mass wasting processes and their control on denudation rates, Matter-Valley, Switzerland
Austria, Vienna, 22.-27.4.2012, EGU General Assembly

F. Kober, K. Hippe, S. Ivy-Ochs, J. Zasadni, R. Wieler, L. Wacker, P.W. Kubik and C. Schlüchter
The Lateglacial deglaciation history of the high Alpine Gotthard Pass, Switzerland, based on cosmogenic ^{10}Be and in situ ^{14}C
Austria, Vienna, 22.-27.4.2012, EGU General Assembly

F. Kober, K. Hippe, B. Salcher, S. Ivy-Ochs, P.W. Kubik, M. Christl and L. Wacker
Debris flows and cosmogenic catchment wide denudation rates
Austria, Vienna, 22.-27.4.2012, EGU General Assembly

F. Kober, K. Hippe, B. Salcher, S. Ivy-Ochs, P.W. Kubik, M. Christl and L. Wacker
Storage of colluvium in alpine headwaters with cosmogenic ^{10}Be and ^{14}C
France, Paris, 09.07.2012, 21st Radiocarbon Conference

F. Kober, K. Hippe, B. Salcher, S. Ivy-Ochs, P.W. Kubik, M. Christl, L. Wacker and N. Hählen
Cosmogenic nuclide denudation rates in the debris-flow dominated Haslital-Aare and Matter catchments
Switzerland, Berne, 16.-17.11.2012, 10th Swiss Geoscience Meeting

M. Kocher, N. Akçar, C. Schlüchter and P.W. Kubik
Unconsolidated sediments on Piz Starlex at >3000m, Swiss-Italian border area
Switzerland, Berne, 10.03.2012, CH-QUAT Annual Meeting 2012

P. Köpfli, J.R. Moore and S. Ivy-Ochs
The Oeschinensee rock avalanche: reconstruction and dating of a prehistoric event
Austria, Vienna, 22.-27.4.2012, EGU General Assembly

M.S. Krzemnicki and I. Hajdas
Radiocarbon ^{14}C dating of pearls
France, Paris, 08.-13.07.2012, 21st Radiocarbon Conference

J. Lachner
AMS of $^{10}Be/^9Be$ and $^{26}Al/^{27}Al$ at low energies
Austria, Vienna, 28.06.2012, VERA Seminar

J. Lachner, M. Christl, A. Müller, H.-A. Synal, M. Schaller and C. Maden
^{10}Be und ^{26}Al Messungen mit Niederenergie AMS
Germany, Stuttgart, 15.03.2012, DPG Spring Meeting

J. Lachner
Carrier freie $^{10}Be/^9Be$ Messungen am Tandy
Switzerland, Zurich, 07.03.2012, AMS Seminar

K. Leith, J. Moore, P. Sternai, S. Ivy-Ochs, F. Hermann and S. Loew
Reconstruction of an extensive Younger Dryas glacial stadial in Canton Valais, Switzerland
Austria, Vienna, 22.-27.4.2012, EGU General Assembly

S. Maxeiner
Umladungswirkungsquerschnitte von ^{12}C in He bei 30-50 keV
Switzerland, Zurich, 14.09.2012, AMS Seminar

A. Michczynski and I. Hajdas
Complex statistical model of the Lake Soppensee chronology
France, Paris, 08.-13.07.2012, 21st Radiocarbon Conference

S.Meyer, J. Leifeld and I. Hajdas
Soil microbial degradation continues after soil organic matter stabilization
Austria, Vienna, 22.-27.4.2012, EGU General Assembly

A.M. Müller, M. Suter, H.-A. Synal, D. Fu, X. Ding, K. Liu and L. Zhou
A simple way of upgrading a compact NEC-radiocarbon AMS facility for ^{10}Be
France, Paris, 09.07.2012, 21nd Radiocarbon Conference

H. Munack, J.H. Blöthe, D. Scherler, H. Wittmann, P.W. Kubik and O. Korup
Postglacial denudation outpaced by long-term exhumation, Ladakh and Zanskar, India
Austria, Vienna, 22.-27.4.2012, EGU General Assembly

R. Reber, N. Akçar, S. Ivy-Ochs, D. Tikhomirov, R. Burkhalter, C. Zahno, A. Lüthold, P.W. Kubik, C. Vockenhuber and C. Schlüchter
The last deglaciation of the northern Alpine Foreland: Evidence from the Reuss Glacier
Switzerland, Berne, 06.-17.11.2012, 10th Swiss Geoscience Meeting

R. Reber, N. Akçar, V. Yavuz, D. Tikhomirov, P.W. Kubik, C. Vockenhuber and C. Schlüchter
Paleoglacier chronology of the southwestern Black Sea region
Switzerland, Berne, 06.-17.11.2012, 10th Swiss Geoscience Meeting

R. Reber, N. Akçar, V. Yavuz, D. Tikhomirov, C. Vockenhuber, P.W. Kubik, C. Schlüchter
Paleoclimate in north-east Anatolia during the Quaternary deduced from glacial archives
Austria, Vienna, 22.-27.4.2012, EGU General Assembly

R. Reber, N. Akçar, S. Ivy-Ochs, D. Tikhomirov, R. Burkhalter, C. Zahno, A. Lüthold, P.W. Kubik, C. Vockenhuber and C. Schlüchter
Deglaciation of the Reuss glacier in the Alps at the end of the Last Glacial Maximum
Switzerland, Berne, 10.03.2012, CH-QUAT Annual Meeting 2012

S. Savi, K.P. Norton, F. Schlunegger, V. Picotti, F. Brardinoni, N. Akçar and P.W. Kubik
[10]Be in the understanding of sediment transfer
Switzerland, Berne, 10.03.2012, CH-QUAT Annual Meeting 2012

S. Savi, K.P. Norton, F. Schlunegger, V. Picotti, F. Brardinoni, N. Akçar and P.W. Kubik
How does sediment mixing affect [10]Be concentrations in alluvial sediments? A case study from a small catchment of the Alps, Zielbach, Alto Adige, Italy
Austria, Vienna, 22.-27.4.2012, EGU General Assembly

S. Savi, K.P. Norton, F. Brardinoni, N. Akçar, P.W. Kubik, V. Picotti and F. Schlunegger
Erosion and sedimentation rate variability following the LGM ice-retreat
Switzerland, Berne, 06.-17.11.2012, 10th Swiss Geoscience Meeting

T.E. Scharf, A.T. Codilean, M. de Wit, J.D. Jansen and P.W. Kubik
Lithologies preserve alpine-like topography in southern Africa
USA, San Francisco, 03.-07.12.2012, American Geophysical Union Fall Meeting

T.E. Scharf, A.T. Codilean, M. de Wit, J.D. Jansen and P.W. Kubik
Lithologies preserve alpine-like topography in southern Africa
Germany, Potsdam, 26.-30.11.2012, GFZ Workshop GEO-FUTURE

D. Scherler, H. Munack, J. Mey, H. Wittmann, P.Kubik and M. Strecker
Prolonged and massive river damming by Siachen Glaciar, Karakoram, during the penultimate glacial period
Austria, Vienna, 22.-27.4.2012, EGU General Assembly

M. Seiler, S. Maxeiner, M. Suter and H.-A. Synal
Helium Strippergas bei Radiocarbon AMS
Germany, Stuttgart, 16.03.2012, DPG Spring Meeting

M. Seiler
myCADAS: First measurements
Switzerland, Zurich, 30.05.2012, AMS Seminar

M. Seiler, S. Maxeiner, M. Suter and H.-A. Synal
Charge exchange cross sections for carbon ions in helium
France, Paris, 09.07.2012, 21st Radiocarbon Conference

M. Seiler, C. Münsterer, F. Rechberger, G. Salazar, L. Wacker and H.-A. Synal
Simple carbon dioxide transport of carbonate samples to a gas ion source
France, Paris, 09.07.2012, 21st Radiocarbon Conference

M.J. Simon, M. Döbeli, A. Eggenberger, A.M. Müller, M. Schulte-Borchers and H.A. Synal
In-air capillary microprobe for STIM and PIXE
Portugal, Lisbon, 22.-27.7.2012, 13th International Conference on Nuclear Microprobe Technology & Applications 2012

M. Suter
Beschleunigermassenspektrometrie (AMS) - von Grossgeräten zu Tischmodellen
Germany, Braunschweig, 12.07.2012, Kolloquium PTB

M. Suter
Challenges in designing compact AMS facilities
New Zealand, Lower Hutt, 19.11.2012, Seminar GNS

H.-A. Synal
Recent advances in ^{14}C Accelerator Mass Spectrometry and its implications for biomedical research
Belgium, Beerse, 26.-27. June 2012, Research Collaboration Meeting

H.-A. Synal, M. Seiler, S. Maxeier and L. Wacker
Progress in mass spectrometric radiocarbon detection techniques
France, Paris, 09.07.2012, 21st Radiocarbon Conference

H.-A. Synal
Reducing size and complexity of accelerator mass spectrometers
Germany, Hannover, 11.07.2012, Festkolloquium des Instituts für Radioökologie und Strahlenschutz

H.-A. Synal
Progress in mass spectrometric ^{14}C detection and its implications for biomedical research
Germany, Heidelberg, 09.-13.September 2012, 11th International Symposium on the Synthesis and Applications of Isotopes and Isotopically Labelled Compounds

H.-A. Synal
Radiokarbon Altersbestimmungen: Neue Nachweismethoden und ausgewählten Anwendungsbeispiele
Switzerland; Zurich; 18.10.2012, Physikalische Gesellschaft Zürich

H.-A. Synal
Latest developments in AMS measurement technologies
Romania, Bucharest-Magurele, 01.11.2012, Workshop on Opportunities for applied research at the new tandem accelerators of IFIN – HH

H.-A. Synal
The art of dating: Progress in radiocarbon detection and related applications
Switzerland, Zurich, 14.11.2012, The Zurich Physics Colloquium

H.-A. Synal
The art of dating - ^{14}C-Altersbestimmungen: Neue Nachweismethoden und ausgewählte Beispiele
Germany, Heidelberg, 07.12.2012, Physikalisches Kolloquium

D. Tikhomirov, N. Akçar, V. Alfimov, S. Ivy-Ochs and C. Schlüchter
Calibration of ^{36}Cl production rate on ^{39}K in Antarctica granites
Switzerland, Berne, 10.03.2012, CH-QUAT Annual Meeting 2012

D. Tikhomirov, N. Akçar, V. Alfimov, S. Ivy-Ochs and C. Schlüchter
Calculation of shielding factors for production of cosmogenic nuclides in fault scarps
Switzerland, Berne, 06.-17.11.2012, 10th Swiss Geoscience Meeting

V. Vanacker, N. Bellin and P.W. Kubik
Constraining long-term denudation rates in the Betic Cordillera, Southern Spain
Italy, Rome, 01.-05.07.2012, IAG Regional Conference

V. Vanacker, N. Bellin, R. Ortega and P.W. Kubik
Human impact on soil erosion rates in SE Spain: an integrated approach
Netherlands, Wageningen, 09.05.2012, Mini-Symposium

V. Vanacker, G. Govers, F. von Blanckenburg and P.W. Kubik
Forest cover change and its effect on soil erosion in tropical mountain regions: an example from the Ecuadorian Andes
Chile, Concepcion, 04.-09.11.2012, IUFRO Landscape Ecology Conference

V. Vanacker, G. Govers, F. von Blanckenburg and P.W. Kubik
Anthropogenic control on geomorphic processes
Germany, Bonn, 20.11.2012, Research seminar

C. Vockenhuber
Optimizing small AMS systems beyond C-14
Germany, Stuttgart, 16.03.2012, DPG Spring Meeting

C. Vockenhuber
AMS within CoDustMas: Nanodiamonds
Switzerland, Ascona, 05.-08.11.2012, Dust in Core-Collapse Supernovae near and far: understanding its formation and evolution

L. Wacker
The MICADAS gas ion source
Belgium, Beerse, 27.06.2012, Research Collaboration Meeting

L. Wacker, M. Christl and H.-A. Synal
Data reduction of small radiocarbon samples
France, Paris, 09.07.2012, 21st Radiocarbon Conference

C. Wirsig, S. Ivy-Ochs, N. Akcar, V. Alfimov, C. Kämpfer and C. Schlüchter
Studying depth of glacial erosion and timing of deglaciation by combining cosmogenic ^{10}Be, ^{14}C and ^{36}Cl
Switzerland, Berne, 10.03.2012, CH-QUAT Annual Meeting 2012

C. Wirsig, S. Ivy-Ochs, N. Akcar, V. Alfimov, C. Kämpfer and C. Schlüchter
Combining cosmogenic Be, C, Al and Cl – Quantifying depth of glacial erosion and timing of deglaciation
Austria, Vienna, 23.04.2012, EGU General Assembly

C. Wirsig and A. Mangini
Mangankruste- Erstellung eines hochaufgelösten Profils
Germany, Heidelberg, 18.06.2012, Institut für Umweltphysik / Paläoklima

C. Wirsig, S. Ivy-Ochs, N. Akcar, H.-A. Synal, R. Wieler and C. Schlüchter
Combining cosmogenic Be, C and Cl – Quantifying depth of glacial erosion and timing of deglaciation
Switzerland, Zurich, 31.08.2012, PhD Proposal Defense

SEMINAR

'CURRENT TOPICS IN ACCELERATOR MASS SPEKTRO-METRY AND RELATED APPLICATIONS'

Spring semester

22.02.2012

Florian Kober (ETHZ, Switzerland), Debris flows and cosmogenically derived denudation rates

29.02.2012

Vasily Alfimov (ETHZ, Switzerland) and Annette Heusser (Oxyphen AG, Wetzikon, Switzerland), Capillary pore membranes for automotive and life science applications

07.03.2012

Johannes Lachner (ETHZ, Switzerland), Carrier-freie ^{10}Be/^{9}Be Messungen am Tandy

14.03.2012

Werner Schoch (Labor für Quartäre Hölzer, Langnau a. A., Switzerland), Holz und Holzkohle - Beispiele aus der Praxis

21.03.2012

Nuria Casacuberta (Universitat Autònoma de Barcelona, Spain), Presence and fate of ^{90}Sr in seawater off Japan as a consequence of the Fukushima Dai-ichi nuclear accident

28.03.2012

Mareike Schwinger (University of Hannover, Germany), Bestimmung von ^{129}I und ^{127}I in Umweltproben

04.04.2012

Jenny Feige (University of Vienna, Austria), The search for supernova-produced radionuclides in new deep-sea sediment samples and preparation for AMS measurements

11.04.2012

Urs Baltensperger (PSI, Switzerland), The CLOUD experiment

18.04.2012

Jens Fohlmeister (University of Heidelberg, Germany), Carbon isotopes in speleothems - Tracers of soil processes

25.04.2012

Yanlin Zhang (University of Bern), Source apportionment of carbonaceous aerosol particles by ^{14}C analysis

02.05.2012

Naki Akcar (University of Bern), The challenge of dating the Swiss Deckenschotter

09.05.2012

Luigi Bruno (University of Bologna, Italy), Subsurface stratigraphy and archeological characterization of the Bologna urban area (Northern Italy)

16.05.2012
Susan Ivy-Ochs (ETHZ, Switzerland), Landslides in Trentino

23.05.2012
Marius Simon (ETHZ, Switzerland), Capillary Ion Probe: Grundlagen
Andreas Eggenberger (ETHZ, Switzerland), Capillary Ion Probe: Auflösungsmessungen

30.05.2012
Martin Seiler (ETHZ, Switzerland), myCADAS - erste Messungen

05.07.2012
Alan Hogg (Waikato Radiocarbon Dating Laboratory, New Zealand), Application of New Zealand sub-fossil Kauri to Radiocarbon calibration

19.07.2012
Keith Fifield (ANU, Australia), Bomb plutonium at the source: adventures with an Enewetak coral

20.08.2012
Nadia El Hachimi (ENSICAEN, Caen, France), The applicability of digital pulse processing to the ETH AMS system

Fall semester

16.09.2012
Sascha Maxeiner (ETHZ, Switzerland), Umladungswirkungsquerschnitte von ^{12}C in He bei 30-50 keV

19.09.2012
Ola Kwiecien (EAWAG, Switzerland), ICDP Paleovan project - 3 months of adventure and 500 ka of environmental changes in eastern Anatolia, Turkey

26.09.2012
Dominique Oppler (LIBRUM, Hochwald, Switzerland), Entdeckung einer metallurgischen Industrie im Hochhimalaja

03.10.2012
Cameron McIntyre (ETHZ, Switzerland), Gas chromatography and radiocarbon-AMS

10.10.2012
Miriam Dünforth (LMU Munich, Germany), Ice on rocks: natural and numerical experiments on glacial erosion

17.10.2012
Gary Salazar (University of Bern), Different CO_2 traps for the MICADAS gas inlet system

24.10.2012
Oliver Forstner (University of Vienna, Austria), The ILIAS project for selective isobar suppression by laser photodetachment

31.10.2012
Anne Claude (ETHZ, Switzerland), Geomorphology and landscape evolution of the Chironico landslide (Leventina valley, southern Swiss Alps)

07.11.2012
Núria Casacuberta (Universitat Autònoma de Barcelona, Spain), Uranium-236 in the North Atlantic

14.11.2012
Feng Xiaojuan (ETHZ, Switzerland), Probing fluvial transfer of organic carbon using lignin phenol ^{14}C ages

21.11.2012
Caroline Münsterer (ETHZ, Switzerland), Laser ablation and radiocarbon-AMS

28.11.2012
Stefan Röllin (Labor Spiez, Spiez, Switzerland), Aktinide in Sedimenten aus schweizer Seen und dem Fluss Yenisei

05.12.2012
Christoph Elsässer (University of Heidelberg, Germany), Disentangling production and climate variability of ice core ^{10}Be: A basic model approach

12.12.2012
Clemens Walther (IRS Hannover, Germany), Das Institut für Radioökologie und Strahlenschutz der Uni Hannover, derzeitige Arbeiten und Perspektiven

19.12.2012
Myriam Krieg (Musée Romain Avenches, Avenches, Switzerland), Bis ans Ende der Patina - Vergleichende Untersuchungen von Grundmetall und Patina archäologischer Kupferlegierungs-Objekte aus Avenches (VD)

THESES (INTERNAL)

Term papers

Andreas Diebold
Measurement of uranium and plutonium in seawater samples by AMS
ETH Zurich

Felix Rechenberger
Optimization of CO_2 transfer to the gas ion source via different trap materials
ETH Zurich

Andreas Eggenberger
Resolution measurements of the capillary microprobe with STIM and PIXE
ETH Zurich

Diploma/Master theses

Sascha Maxeiner
Kohlenstoffstripping in Helium bei 45 keV
ETH Zurich

THESES (EXTERNAL)

Diploma/Master theses

Benjamin Campforts
Explain river response in the Ecuadorian Andes: Reconciling spatial and temporal scales
KULeuven (Belgium)

Anne Claude
Geomorphology and landscape evolution of the Chironico landslide (Leventina valley, southern Swiss Alps)
ETH Zurich (Switzerland)

Anja Cording
Climatic and tectonic implications of Pleistocene river incision derived from ^{10}Be exposure dating of depositional surfaces at the Dzungarian Alatau, northern Tien Shan
University of Münster (Germany)

Armin Dielforder
The deglaciation history of the Simplon Pass area derived from ^{10}Be exposure dating of ice-molded bedrock surfaces
University of Münster (Germany)

Nadia El Hachimi
Data acquisition and processing of nuclear events with the CAEN Desktop Digitizer
ENSICAEN (France)

Loïc Fave
Determination of helium implantation induced swelling in ODS steels
EPF Lausanne (Switzerland)

Lorenz Grämiger
Prehistoric rock avalanches at Rinderhorn, Switzerland
ETH Zurich (Switzerland)

Mario Kocher
Die Lockergesteine am Piz Starlex
University of Bern (Switzerland)

Patrizia Köpfli
The Oeschinensee Rock Avalanche: Reconstruction and Dating
ETH Zürich (Switzerland)

Taryn E. Scharf
Denudation rates and geomorphic evolution of the Cape Mountains, determined by the analysis of in situ-produced cosmogenic Be-10
University of Cape Town (South Africa)

Andreas Struffert
10Be-derived denudation rates in the central Menderes Massif, southwestern Turkey
University of Münster (Germany)

Doctoral theses

Debajeet Bora
Hematite and its hybrid nanostructures for photoelectrochemical water splitting
University of Basel (Switzerland)

Henning Galinski
Metals on ceramics : agglomeration, nano-corrosion, growth and oxidation
ETH Zurich (Switzerland)

Kristina Hippe
Combining cosmogenic ^{10}Be and in situ ^{14}C in Earth surface sciences: A new ^{14}C extraction system and two case studies on sediment transfer and surface exposure dating
ETH Zurich (Switzerland)

Dominik Jaeger
Interface investigations on titanium nitride bilayer systems
EPF Lausanne (Switzerland)

Kerry Leith
Stress development and geomechanical controls on the geomorphic evolution of Alpine valleys
ETH Zurich (Switzerland)

Julian Perrenoud
Low temperature grown CdTe thin film solar cells for the application on flexible substrates
ETH Zurich (Switzerland)

Bernhard Schwanitz
Reduzierung der Platinbeladung und Imaging von Alterungsphänomenen in der Polymerelektrolyt-Brennstoffzelle
ETH Zurich (Switzerland)

Matteo Seita
Full microstructure control through ion-induced grain growth, texturing and constrained deformation in thin films
ETH Zurich (Switzerland)

Akshath Shetty
Off-axis texture and crystallographic accommodation in multicomponent nitride thin films deposited by pulsed magnetron sputtering
EPF Lausanne (Switzerland)

Marcus Strobl
Landscape evolution of an exceptionally well preserved bedrock peneplain on the southeastern Tibetan Plateau
University of Münster (Germany)

Fabian Van den Berg
Erosion and sediment transport in the Entlen River, Switzerland
University of Bern (Switzerland)

COLLABORATIONS

Australia

The Australian National University, Department of Nuclear Physics, Canberra

Austria

AlpS - Zentrum für Naturgefahren- und Riskomanagement GmbH, Geology and Mass Movements, Innsbruck

Geological Survey of Austria, Sediment Geology, Vienna

University of Innsbruck, Institute of Botany, Innsbruck

University of Innsbruck, Institute of Geography, Innsbruck

University of Innsbruck, Institute of Geology, Innsbruck

University of Vienna, VERA, Faculty of Physics, Vienna

Vienna University of Technology, Institute for Geology, Vienna

Belgium

Royal Institute for Cultural Heritage, Brussels

University of Louvain, Department of Geography, Louvain

Canada

Chalk River Laboratories, Dosimetry Services, Ottawa

TRIUMF, Vancouver

China

China Earthquake Administration, Beijing

Peking University, Accelerator Mass Spectrometry Laboratory, Beijing

Peking University, Department of Geography, College of Urban and Environmental Sciences, Beijing

Denmark

Danfysik, A/S, Taastrup

Risø DTU, Risø National Laboratory for Sustainable Energy, Roskilde

University of Southern Denmark, Department of Physics, Chemistry and Pharmacy, Odense

Finnland

University of Jyväskylä, Physics Department, Jyväskylä

France

Aix-Marseille University, Collège de France, Aix-en-Provence

Laboratoire de Biogeochimie Moléculaire, Strasbourg

Université de Savoie, Laboratoire EDYTEM, Le Bourget du Lac

Université Paris Panthéon-Sorbonne, Laboratoire de Géographie Physique, Meudon cedex

Germany

Alfred Wegener Institute of Polar and Marine Research, Marine Geochemistry, Bremerhaven

BSH Hamburg, Radionuclide Section, Hamburg

Deutsches Bergbau Museum, Bochum

Diözesanmuseum Freising, Erzbischöfliches Ordinariat München, Freising

German Research Centre for Geosciences (GFZ), Potsdam

Helmholtz Zentrum Berlin, Heterogeneous Materials Systems, Berlin

Hydroisotop GmbH, Schweitenkirchen

IFM-GEOMAR, Palaeo-Oceanography, Kiel

Regierungspräsidium Stuttgart, Landesamt für Denkmalpflege, Esslingen

University of Bremen, Geosciences, Bremen

University of Cologne, Theoretical Chemistry, Cologne

University of Cologne, Institute of Geology and Mineralogy, Cologne

University of Freiburg, Institut für Vorderasiatische Archäologie, Freiburg

University of Hannover, Center for Radiation Protection and Radioecology, Hannover

University of Hannover, Institute of Geology, Hannover

University of Heidelberg, Institute of Environmental Physics, Heidelberg

University of Münster, Institute of Geology and Paleontology, Münster

University of Potsdam, Institute for Geosciences, Potsdam

University of Tübingen, Department of Geosciences, Tübingen

Direktion Landesarchäologie, Speyer

Hungary

Hungarian Academy of Science, Institute of Nuclear Research (ATOMKI), Debrecen

India

Inter-University Accelerator Center, Accelerator Division, New Dehli

Israel

Hebrew University, Geophysical Institute of Israel, Jerusalem

Italy

Geological Survey of the Provincia Autonoma di Trento, Landslide Monitoring, Trentino

Istituto Nazionale di Geofisica e Vulcanologia, Sez. Sismologia e Tettonofisica, Rome

University of Bologna, Deptartment Earth Sciences, Bologna

University of Padua, Department of Geology and Geophysics, Padua

Universty of Turin, Department of Earth Sciences, Turin

University of Salento, Department of Physics, Lecce

Lichtenstein

OC Oerlikon AG, Balzers

Mexico

UNAM (Universidad Nacional Autonoma de Mexico), Instituto de Fisica, Mexico

New Zealand

Victoria University of Wellington, School of Geography, Environment and Earth Sciences, Wellington

Norway

Norwegian University of Life Sciences, Department of Plant and Environmental Sciences, Ås

The University Centre of Svalbard, Quaternary Geology, Longyearbyen

University of Bergen, Department of Earth Science, Bergen

University of Bergen, Department of Biology, Bergen

University of Norway, The Bjerkness Centre for Climate Res., Bergen

Romania

Horia Hulubei - National Institute for Physics and Nuclear Engineering, Magurele

Russia

Russian Academy of Sciences, Laboratory of Ion and Molecular Physics of the Institute for Energy Problems of Chemical Physics, Moscow

Singapore

National University of Singapore, Department of Chemistry, Singapore

Slovakia

Comenius University, Faculty of Mathematics, Physics and Infomatics, Bratislava

South Africa

Nelson Mandela Metropolitan University, Port Elizabeth

Spain

University of Seville, National Center for Accelerators, Seville

University of Murcia, Department of Plant Biology, Murcia

Universitat Autònoma de Barcelona, Environmental Science and Technology Institute, Barcelona

Sweden

Lund University, Department of Earth and Ecosystem Sciences, Lund

Onsala Space Obervatory, Onsala

University of Upsalla, Angström Institute, Upsalla

University of Upsalla, Tandem Laboratory Upsalla

Switzerland

ABB Semiconductors AG, Baden

Bopp AG, Zurich

Dectris AG, Baden

Dendrolabor Wallis, Brig

Empa, Functional Polymers, Dübendorf

Empa, Joining Technology and Corrosion, Dübendorf

Empa, Advanced Materials Processing, Thun

Empa, High Performance Ceramics, Dübendorf

Empa, Nanoscale Materials, Dübendorf

Empa, Thin Films, Dübendorf

Empa, Mechanics of Materials and Nanostructures, Dübendorf

ENSI, Brugg

EPFL Lausanne, Institute de Physique de la Matière Condensée, Lausanne

ETH Zurich, Engineering Geology, Zurich

ETH Zurich, Institute of Food Nutrition and Health, Zurich

ETH Zurich, Institute of Geochemistry and Petrology, Zurich

ETH Zurich, Institute of Geology, Zurich

ETH Zurich, Institute of Isotope Geochemistry and Mineral Resources, Zurich

ETH Zurich, Laboratory of Inorganic Chemistry, Zurich

ETH Zurich, Metals Research, MATL, Zurich

ETH Zurich, Nonmetallic Materials, MATL, Zurich

ETH Zurich, Solid State Physics, Zurich

ETH Zurich, Micro- and Nanosystems, MAVT, Zurich

ETH Zurich, Optical Materials Engineering, MAVT, Zurich

Evatec, Flums

Gübelin Gem Lab Ltd. (GGL), Luzern

Haute Ecole ARC, IONLAB, La-Chaux-de-Fonds

Helmut Fischer AG, Hünenberg

Inficon AG, Balzers

Kanton Bern, Achäologischer Dienst, Berne

Kanton Graubünden, Archäologischer Dienst, Chur

Kanton Solothurn, Kantonsarchäologie, Solothurn

Kanton St. Gallen, Kantonsarchäologie, St. Gallen

Kanton Zug, Kantonsarchäologie, Zug

Kanton Zürich, Kantonsarchäologie, Dübendorf

Labor für quartäre Hölzer, Affoltern a. Albis

Laboratoire Romand de Dendrochronologie, Moudon

Office et Musée d'Archéologie Neuchatel, Neuchatel

Oxyphen AG, Research and Development, Wetzikon

Paul Scherrer Institut (PSI), Laboratory for Atmospheric Chemistry, Villigen

Paul Scherrer Institut (PSI), Laboratory for Radiochemistry and Environmental Chemistry, Villigen

Paul Scherrer Institut (PSI), Materials Group, Villigen

Paul Scherrer Institut (PSI), Nuclear Materials, Villigen

Paul Scherrer Institut (PSI), Muon Spin Rotation, Villigen

Research Station Agroscope Reckenholz-Tänikon ART, Air Pollution / Climate Group, Zurich

Stadt Zürich, Amt für Städtebau, Zurich

Swiss Federal Institute for Forest, Snow and Landscape Reseach (WSL), Landscape Dynamics, Dendroecology, Birmensdorf

Swiss Federal Institute for Forest, Snow and Landscape Reseach (WSL), Soil Sciences, Birmensdorf

Swiss Federal Institute of Aquatic Science and Technology (Eawag), SURF, Dübendorf

Universität Basel, Departement Altertumswissenschaften, Basel

University of Basel, Department of Physics, Basel

University of Basel, Institut für Prähistorische und Naturwissenschaftliche Archäologie (IPNA), Basel

University of Bern, Institute of Geology, Berne

University of Bern, Oeschger Center for Climate Research, Berne

University of Bern, Climate and Environmental Physics, Berne

University of Bern, Department of Chemie and Biochemistry, Berne

University of Fribourg, Department of Physics, Fribourg

University of Geneva, Department of Anthropology and Ecology, Geneva

University of Geneva, Department of Geology and Paleontology, Geneva

University of Lausanne, Institute of Geomatics and Risk Analysis, Lausanne

University of Neuchatel, Department of Geology, Neuchatel

University of Zurich, Abteilung Ur- und Frühgeschichte, Zurich

University of Zurich, Institute of Geography, Zurich

University of Zurich, Paläontologisches Institut und Museum, Zurich

Turkey

Istanbul Technical University, Faculty of Mines, Istanbul

United Kingdom

Durham University, Department of Geography, Durham

University of Bristol, Department of Earth Sciences, Bristol

University of Manchester, Corrosion Protection Center, Manchester

USA

Columbia University, LDEO, New York

Eckert & Ziegler Vitalea Science, AMS Laboratory, Davis

Idaho National Laboratory, National and Homeland Security, Idaho Falls

University of Utah, Geology and Geophysics, Salt Lake City

VISITORS AT THE LABORATORY

Valentina Gaballo
CEnter for DAting e Diagnostics (CEDAD), University of Salento, Brindisi, Italy
16.01.2012 - 30.04.2012

Mathias Bichler
Department of Environmental Geosciences, University of Vienna, Vienna, Austria
01.02.2012 - 29.02.2012

Martin Reindl
Department of Environmental Geosciences, University of Vienna, Vienna, Austria
01.02.2012 - 29.02.2012

Luigi Bruno
Department of Biological, Geological and Environmental Sciences, University of Bologna, Bologna, Italy
20.02.2012 - 31.05.2012

Gary Salazar Quintero
Department of Chemistry & Biochemistry, University of Bern, Switzerland
20.02.2012 - 30.09.2012

Corina Solis Rosales
Institute of Physics, Universidad Nacional Autonoma de Mexico, Mexico
01.03.2012 - 31.08.2012

Jörg Schäfer
LDEO, Columbia University, New York, USA
02.03.2012 - 02.03.2012

Santiago Padilla
Atomic, Molecular and Nuclear Physics, University of Seville, Spain
15.03.2012 - 15.09.2012

Nuria Casacuberta
Department of Physics, Universitat Autònoma de Barcelona, Spain
18.03.2012 - 06.04.2012

Toni Wallner
Department of Nuclear Physics, Australian National University, Canberra, Australia
22.03.2012 - 23.03.2012

Mareike Schwinger
Institute for Radioecology and Radiation Protection, Leibniz University Hannover, Hannover, Germany
26.03.2012 - 28.03.2012

Javier Santos
Centro Nacional de Aceleradores, University of Seville, Spain
10.04.2012 - 20.04.2012

Jong Han Song
Korea Institute of Science and Technology, Seoul, South Korea
23.04.2012 - 23.04.2012

Joon Kon Kim
Korea Institute of Science and Technology, Seoul, South Korea
23.04.2012 - 23.04.2012

Jae Yeoul Kim
Korea Institute of Science and Technology, Seoul, South Korea
23.04.2012 - 23.04.2012

Byung Yong Yu
Korea Institute of Science and Technology, Seoul, South Korea
23.04.2012 - 23.04.2012

Andrew P. Moran
Institute for Geography, University of Innsbruck, Austria
01.05.2012 - 31.05.2012

Nadia El Hachimi
L'Ecole Nationale Supérieure d'Ingéniers de Caen et Centre de Recherche, Caen, France
02.05.2012 - 20.08.2012

Edouard Bard
Laboratory of Environmental Isotopic Geochemistry, Université d'Aix Marseille, Aix en Provence, France
16.05.2012 - 16.05.2012

Irene Schimmelpfennig
Laboratory of Environmental Isotopic Geochemistry, Université d'Aix Marseille, Aix en Provence, France
16.05.2012 - 16.05.2012

Amandine Perret
Laboratoire EDYTEM, Université de Savoie, Le Bourget du Lac, France
04.06.2012 - 22.06.2012

Marco De Zorzi
Department of Geosciences, University of Padova, Padova, Italy
10.06.2012 - 17.07.2012

Nuria Casacuberta
Department of Physics, Universitat Autònoma de Barcelona, Spain
11.06.2012 - 11.08.2012

Xavier Maeder
CSEM Neuchâtel, Neuchâtel, Switzerland
18.06.2012 - 20.06.2012

Alfred Dewald
Institute of Nuclear Physics, University of Cologne, Cologne, Germany
21.06.2012 - 22.06.2012

Stefan Heinze
Institute of Nuclear Physics, University of Cologne, Cologne, Germany
21.06.2012 - 22.06.2012

Keith Fifield
Department of Physics, Australian National University, Canberra, Australia
19.07.2012 - 24.07.2012

Tobias Lutz
AKAD, Zürich, Switzerland
23.08.2012 - 27.08.2012

Dominique Oppler
IPNA, University of Basel, Switzerland
23.08.2012 - 27.08.2012

Dongpo Fu
Institute of Heavy Ion Physics, Peking University, Beijing, China
01.09.2012 - 23.09.2012

Stefan Bister
Institute for Radioecology and Radiation Protection, Leibniz University Hannover, Hannover, Germany
04.09.2012 - 05.09.2012

Viktoria Schauer
Atominstitut, Vienna, Austria
04.09.2012 - 05.09.2012

Stefanie Schneider
Institute for Radioecology and Radiation Protection, Leibniz University Hannover, Hannover, Germany
04.09.2012 - 05.09.2012

Nicoleta Florea
National Institute of Physics and Nuclear Engineering, Bucharest, Romania
13.09.2012 - 20.09.2012

Corina Anca Simion
National Institute of Physics and Nuclear Engineering, Bucharest, Romania
13.09.2012 - 20.09.2012

Pia Sköld
Department of Geology, Lund University, Sweden
13.09.2012 - 20.09.2012

Dhafira Benzeggouta
Paris NanoSciences Institute (INSP), Université de Pierre et Marie Curie, Paris, France
24.09.2012 - 24.09.2012

Daniel Hanf
Helmoltz-Zentrum Dresden-Rossendorf, Rossendorf, Germany
24.09.2012 - 25.09.2012

Ian Vickridge
Paris NanoSciences Institute (INSP), Université de Pierre et Marie Curie, Paris, France
24.09.2012 - 24.09.2012

Marco Stucki
Kantonsschule Aarau, Menziken, Switzerland
15.10.2012 - 19.10.2012

Mathieu Boudin
Royal Institute for Cultural Heritage, Brussels, Belgium
21.10.2012 - 27.10.2012

Tess Van den Brande
Royal Institute for Cultural Heritage, Brussels, Belgium
21.10.2012 - 27.10.2012

Bruno Luigi
Department of Biological, Geological and Environmental Sciences, Università degli Studi di Bologna, Italy
22.10.2012 - 31.10.2012

Nuria Casacuberta
Department of Physics, Universitat Autònoma de Barcelona, Spain
01.11.2012 - 07.11.2012

Mathieu Boudin
Royal Institute for Cultural Heritage, Brussels, Belgium
05.11.2012 - 09.11.2012

Tess Van den Brande
Royal Institute for Cultural Heritage, Brussels, Belgium
05.11.2012 - 09.11.2012

Mark Van Strydonck
Royal Institute for Cultural Heritage, Brussels, Belgium
05.11.2012 - 09.11.2012

Björn Dittmann
Institute of Nuclear Chemistry, University of Cologne, Cologne, Germany
07.11.2012 - 08.11.2012

Eren Cihan Karsu
Department of Physics, Celal Bayar University, Manisa, Turkey
08.11.2012 - 16.11.2012

Heinz Gäggeler
Paul Scherrer Institut (and University of Bern), Villigen, Switzerland
09.11.2012 - 09.11.2012

Stefano Pucci
Istituto Nazionale di Geofisica e Vulcanologia, Rome, Italy
12.11.2012 - 02.12.2012

Timon Willi
Kantonsschule Olten, Switzerland
19.11.2012 - 23.11.2012

Cyrill Burgener
Kantonsschule Olten, Switzerland
19.11.2012 - 23.11.2012

Mario Burger
Spiez Laboratory, Spiez, Switzerland
28.11.2012 - 28.11.2012

Stefan Röllin
Spiez Laboratory, Spiez, Switzerland
28.11.2012 - 28.11.2012

Christoph Elsässer
Institute of Environmental Physics, University of Heidelberg, Heidelberg, Germany
05.12.2012 - 05.12.2012

Dai Xiongxin
Chalk River Lalboratories, Atomic Energy of Canada Ltd., Ontario, Canada
06.12.2012 - 06.12.2012

Manuel Raiwa
Institute for Radioecology and Radiation Protection, Leibniz University Hannover, Hannover, Germany
10.12.2012 - 13.12.2012

Mareike Schwinger
Institute for Radioecology and Radiation Protection, Leibniz University Hannover, Hannover, Germany
10.12.2012 - 13.12.2012

Silvana Martin
Geoscience Department, University of Padua, Italy
18.12.2012 - 20.12.2012